山东省自然科学基金（No.ZR2019MA017）、
泰山学院学术著作出版基金研究成果

分数阶非线性振动系统动力学

国忠金　杨洪祥　著

吉林大学出版社

长春

图书在版编目（CIP）数据

分数阶非线性振动系统动力学 / 国忠金，杨洪祥著. —
长春：吉林大学出版社，2021.6
ISBN 978-7-5692-8404-1

Ⅰ.①分… Ⅱ.①国… ②杨… Ⅲ.①非线性振动－
动力系统（数学）Ⅳ.①O194

中国版本图书馆CIP数据核字（2021）第 111199 号

书　　名　　分数阶非线性振动系统动力学
　　　　　　FENSHUJIE FEIXIANXING ZHENDONG XITONG DONGLIXUE

作　　者　　国忠金　杨洪祥　著
策划编辑　　杨占星
责任编辑　　赵洪波
责任校对　　张文涛
装帧设计　　文　一
出版发行　　吉林大学出版社
社　　址　　长春市人民大街 4059 号
邮政编码　　130021
发行电话　　0431－89580028/29/21
网　　址　　http://www.jlup.com.cn
电子邮箱　　jdcbs@jlu.edu.cn
印　　刷　　河北华商印刷有限公司
开　　本　　787mm×1092mm　1/16
印　　张　　11.5
字　　数　　200 千字
版　　次　　2021 年 6 月　第 1 版
印　　次　　2021 年 6 月　第 1 次
书　　号　　ISBN 978-7-5692-8404-1
定　　价　　59.80 元

PREFACE

As we all know, the exact solutions to most nonlinear dynamical equations are unavailable. Even where an exact solution to a nonlinear equation is available, the resulting expression is presented in terms of implicit functions. Furthermore, numerical integration methods cannot provide an overall view of the behavior of the systems in response and dynamics. The quest for accurately approximate solutions to nonlinear dynamic problems in the fields of engineering, applied mathematics, physical sciences and social sciences have led to the development of many analytical or semi-analytical techniques. The focus of this dissertation is on the steady state responses of the fractional ordinary differential equation (FODE) by constructing a new approximate method for quantitative analysis.

When the steady state periodic solutions are concerned, the classical harmonic balance has well known advantages in comparison with time-domain steady state simulation. However, though it is well suited for weakly and mildly nonlinear problems, it has essential limitations for analysis of strongly nonlinear systems. To obtain desired accuracy highly nonlinear problem requires a large number of harmonics, which may increase the difficulty of the system to be solved. In this thesis, based on the classical harmonic balance method and homotopy technique, residue harmonic balance and method of residue harmonic balance in combination with polynomial homotopy continuation are developed and elaborated for the first time. On the one hand, by bringing homotopy method into the harmonic balance procedure to construct a new approximate technique and balance all the harmonic residues, which is called the residue harmonic balance method, analytical approximations to the nonlinear oscillation of certain autonomous systems are obtained to any desired accuracy. The residue is separated in two parts in each step of residue harmonic balance procedure, one confirms with the present order of approximation and the remaining part for use in the next approximation. The corrections are governed by a set of linear differential equations including harmonic residues that can be easily solved. The residue harmonic balance is introduced to overcome the difficulty of achieving higher-order analytical approximations by using the classical harmonic balance method. On the other hand, polynomial homotopy continuation technique is introduced to obtain all the possible multiple solutions of harmonic balance equations in steady state so that the bifurcation analysis can be implemented. Homotopy continuation methods have proven to be reliable for computing numerically approximations to all isolated solutions of polynomial systems.

The demands of modern technology require a certain revision of the well-established pure mathematical approach. There have appeared a number of works where fractional derivatives are used for a better description of material properties. This book mainly studies the bifurcation and resonance behavior of fractional order systems, focusing on the dynamic behavior of fractional order systems by using the residue harmonic balance method developed by the Editor. The book is divided into five chapters. Chapter 1 introduces two new prosperous analytical methods based on harmonic balance, namely residue harmonic balance and polynomial homotopy continuation techniques. Chapter 2 mainly studies the asymptotic analysis of autonomous fractional ordinary differential systems about the fractional Van der Pol oscillator and autonomous Duffing-Van der Pol oscillator with fractional derivative. Chapter 3 focuses on the research about the resonance response and bifurcation of periodically excited systems containing fractional derivative. The resonance response and analytical approximation of a Duffing oscillator with fractional-order derivative as well as the transition curves and bifurcations of a class of fractional Mathieu-type equations are investigated in detail. Chapter 4 considers the resonance response and effects of fractional differentiation orders on coupled vibration systems. The effects of fractional differentiation orders on a coupled Duffing circuit as well as hard-spring bistability and effect of system parameters in a two-degree-of-freedom vibration system with damping modelled by a fractional derivative are discussed based on residue harmonic balance technique as mentioned above. Some abundant nonlinear dynamical phenomena are found in this section. Chapter 5 focuses on the applications of fractional derivate to structural engineering.

The related research contents and publication of this book are supported by the natural science foundation of Shandong Province (No. ZR2019MA017) and the academic works publishing fund of Taishan University. I would like to express my sincere thanks. Due to the short writing time of this book and the limited level of the author, it is inevitable that there are loopholes and deficiencies. We sincerely invite experts and readers to criticize and correct, and we welcome the exchange and discussion of academic issues.

Guo Zhongjin

March 2021 in Tai shan

TABLE OF CONTENTS

Chapter 1 Residue Harmonic Balance and Polynomial Homotopy Continuation Techniques

1.1 Introduction

Using Fourier series to analyze the time periodic solutions of nonlinear oscillators is well- known. The harmonic balance method, as a powerful tool, consists of first substituting a truncated Fourier expansion into the governing equations. Then by using the orthogonal properties of the trigonometric functions, the coefficients of the harmonic terms are equated to zero and the resulting system of nonlinear algebraic equations is solved for the unknowns. Importantly, harmonic balance is not only valid for both weakly and strongly nonlinear oscillators, but also provides uniformly valid solutions in the whole solution domain for many crucial systems. In general, the first-order harmonic balance method, including one harmonic term in the analysis, is easy to apply to the system with polynomial nonlinearities with a dominant first harmonic. However, more harmonic terms have to include in the analysis for certain motions, especially strongly nonlinear system. As a result, it may be very difficult or impossible to implement the classical harmonic balance method.

In the past few decades, many improved harmonic balance methods have been well developed and employed successfully to study some nonlinear dynamical systems. For instance, Lau [S.L. Lau, 1982] proposed the incremental harmonic balance method that linearizes the set of nonlinear ordinary differential equations along an incremental path to get a set of the linear algebraic equations at a time. Therefore, bifurcation analysis becomes difficult because linear equations would not give multiple solutions. Lim and Wu [C.W. Lim et al., 2002, 2003, 2006, 2007,

2009] proposed linearized harmonic balance (LHB) method by combining the linearization of the governing equation with the harmonic balance method, and Newton harmonic balance (NHB) method that couples of Newton's method with the harmonic balance method. These two methods were successfully used to study certain autonomous ordinary differential systems. However, the above studies adopt simple initial conditions and neglect some harmonic balance residuals that seem to introduce new limitations. Recently, Hall et al [K.C. Hall et al., 2002] and Beran et al [S.P. Beran et al., 2004] developed a high dimensional harmonic balance method which suitable for high dimensional dynamical systems. The key aspect of the HDHB method is that instead of working in terms of Fourier coefficient variables as in the conventional HB approach, the dependent variables are instead cast in the time domain and stored at many equally spaced sub-time levels over the period of one cycle of motion. This makes the HDHB method easy to formulate within the framework of an existing time marching nonlinear solver. However, the HDHB method may produce non-physical solutions as well as the physically meaningful ones. Besides, the number of harmonics retained in the HDHB procedure usually is bigger and even double or higher times, compared to that included in the HB method for the same accuracy of the results. In an attempt to improve the shortcomings of the existing methods for the analysis of nonlinear dynamic systems, a new approach, entitled residue harmonic balance, is presented for the first time in the following section.

1.2 Residue Harmonic Balance Method

When seeking a solution in series form, the number of terms to get a satisfactory solution is not usually known in the beginning. The solution process is required to repeat if the number of terms changes. It is particularly difficult for nonlinear problems. A hierarchical formulation is proposed so that when an approximation is available, the increase of the number of terms constitutes the residual of the previous solution and a set of linear equations associated with corrections for all the existing and new terms is resulted. The solution of the linear equations is rather straight forwards. When applying to harmonic balance for periodic solutions, we propose a novel approximate method, which is termed the residue harmonic balance (RHB). Without loss of generality, here we illustrate the solution procedure of the method corresponding to second ODE system.

1.2.1 Autonomous Differential Equation

Consider a second order nonlinear ordinary differential system of the form

$$\Phi(\ddot{u},\dot{u},u,\lambda) = 0 \tag{1.1}$$

with initial condition $u(0) = A, \dot{u}(0) = 0$, where an over dot denotes differentiation with respect to time t; u stands for the dimensionless unknown functions of time, λ is a control parameter.

Suppose the angular frequency of the Eq. (1.1) is ω which is an unknown that to be determined. Introduce a new time variable τ,

$$\tau = \omega t \tag{1.2}$$

Then, Eq. (1.1) can be rewritten as a system of period 2π

$$\Phi(\omega^2 u'', \omega u', u, \lambda) = 0 \tag{1.3}$$

where a prime denotes differentiation with respect to τ.

Without loss of generality, ones assume that the limit cycle of Eq. (1.3) are symmetrically periodic motions with the angular frequency ω, thus the unknown function u can be expressed by Fourier functions so that

$$u(\tau) = \sum_{k=0}^{\infty} a_{2k+1} \cos[(2k+1)\tau] + \sum_{k=1}^{\infty} b_{2k+1} \sin[(2k+1)\tau] \tag{1.4}$$

Where $a_1, a_{2k+1}, b_{2k+1} (k = 1, 2, \cdots)$ are singly subscripted unknown coefficients to be determined. Obviously, when Eq. (1.1) is conservative, the coefficients of sine terms $b_{2k+1} = 0 (k = 1, 2, \cdots)$.

According to initial condition, one assumes

$$u_0(\tau) = a_{1,0}\cos\tau, \quad \omega = \omega_0 \tag{1.5}$$

be the initial approximation of the solution $u(\tau)$ and the angular frequency ω, respectively. The second subscript 0 of the coefficient a denotes the initial approximation. When Eq. (1.1) is a conservative system, the steady state amplitude is dependent on initial conditions and we have $a_{1,0} = A$. When Eq. (1.1) is a self-excited system, the limit cycle topology is independent by the initial conditions. Thus, the steady state $a_{1,0}$ is determined by system.

On basis of ideas of homotopy perturbation, introducing an order parameter p with values in the interval [0, 1] and rewriting $u(\tau)$ to $u(\tau, p)$, and ω to $\omega(p)$, where $u(\tau, p)$ and $\omega(p)$ have continuous derivatives with respect to p. We expand the approximate solutions in the mth -order truncated Maclaurin series about p,

$$u(\tau, p) = \sum_{k=0}^{m} u_k(\tau)p^k, \quad \omega(p) = \sum_{k=0}^{m} \omega_k p^k \tag{1.6}$$

The Fourier expansion of $u_k(\tau)$ has the similar forms as Eq. (1.4)

$$u_k(\tau) = \sum_{i=0}^{k} a_{2i+1,k}\cos[(2i+1)\tau] + \sum_{i=1}^{k} b_{2i+1,k}\sin[(2i+1)\tau] \tag{1.7}$$

where the second subscripts of a and b denote the order of corrections so that at $p = 1$, the mth-order approximate solutions can be described as

$$u_{(m)}(\tau) = \sum_{k=0}^{m} u_k(\tau), \quad \omega_{(m)} = \sum_{k=0}^{m} \omega_k \tag{1.8}$$

The following is devoted to solve for $u_{(m)}$ and $\omega_{(m)}$ sequentially.

For the zeroth-order approximation ($m = 0$), substituting Eq. (1.5) into Eq. (1.3), yields

$$\Phi(\omega_0^2 u_0'', \omega_0 u_0', u_0, \lambda) = R_u^0(\tau)$$

where the residue $R_u^0(\tau)$ should be zero for an exact approximation but not zero in general. Introducing the order parameter p and rewriting $R_u^0(\tau)$ to $R_u^0(\tau, p)$ and putting them as Fourier series, one has

$$R_u^0(\tau, p) = \Gamma_{1,0}^c \cos\tau + \Gamma_{1,0}^s \sin\tau + \sum_{i=1}^{(\kappa-1)/2} p^i \{\Gamma_{2i+1,0}^c \cos[(2i+1)\tau]$$
$$+\Gamma_{2i+1,0}^s \sin[(2i+1)\tau]\}$$

(1.9)

where $\Gamma_{2i+1,0}^c$ and $\Gamma_{2i+1,0}^s$ are coefficients to be determined. κ is an odd integer depending on the nature of the nonlinearity for symmetric system. We separate the Fourier series in Eq. (1.9) into two parts. The first part has the same number of Fourier terms as the initial approximation in Eq. (1.5). For linear problem, $\kappa=1$, i.e., the second part vanishes. When the nonlinearity is polynomial with the highest power of the product $\ddot{u}^{\kappa_1}\dot{u}^{\kappa_2}u^{\kappa_3}$, then $\kappa=\kappa_1+\kappa_2+\kappa_3$ for integers κ_1,κ_2 and κ_3. For rational or irrational powers, $\ddot{u}^{\kappa_1}\dot{u}^{\kappa_2}u^{\kappa_3}$ will be expanded in Fourier series of infinite number of terms, κ is infinity to be exact.

The right hand side of Eq. (1.9) should not contain secular terms of $\sin(\tau)$ and $\cos(\tau)$. Letting their coefficients be zeros, one has

$$\Gamma_{1,0}^c(a_{1,0},\omega_0) = \frac{1}{\pi}\int_0^{2\pi} R_u^0(\tau)\cos(\tau)\mathrm{d}\tau = 0$$
$$\Gamma_{1,0}^s(a_{1,0},\omega_0) = \frac{1}{\pi}\int_0^{2\pi} R_u^0(\tau)\sin(\tau)\mathrm{d}\tau = 0$$

(1.10)

There are two unknowns $a_{1,0}$ and ω_0 to be determined by Eq. (1.10) which contains two equations. Thus, the zeroth-order approximation to angular frequency and amplitude can be obtained by solving Eq. (1.10). Obviously, when Eq. (1.1) is a conservative system, the coefficient of $\sin(\tau)$ becomes zero, the Eq. (1.10) contains only one equation which is used to determine the angular frequency and amplitude $a_{1,0}$ is initial condition A.

Owing to the nonlinearity of Eq. (1.3), substituting the obtained zeroth-order approximation into Eq. (1.3), Eq. (1.9) is in general nonzero when $\kappa>1$ and the second part of Eq. (1.9) cannot be identically zero. It will inadvertently appear the residuals of the neglected higher harmonics. In fact, these residuals imply the accuracy of approximate solutions. In the following, we construct higher-order analytical approximations based on the remaining residuals in the second part of

Eq. (1.9).

To find the first-order approximation ($m=1$), substituting Eqs. (1.6), (1.7) and (1.9) into Eq. (1.3) and equating the coefficients of the p, we can get

$$\left(\omega_1\frac{\partial}{\partial\omega}+u_1''\frac{\partial}{\partial u''}+u_1'\frac{\partial}{\partial u'}+u_1\frac{\partial}{\partial u}\right)\Phi_0+\Gamma_{3,0}^c\cos(3\tau)+\Gamma_{3,0}^s\sin(3\tau)=R_u^1(\tau)\tag{1.11}$$

where $\partial\Phi_0/\partial u$ denotes that $\partial\Phi/\partial u$ is to be evaluated at the initial approximation after differentiation etc. It is noted that Eq. (1.11) is linear with respect to u_1 and ω_1. Rewrite $R_u^1(\tau)$ to $R_u^1(\tau,p)$, and expand them in Fourier series

$$R_u^1(\tau,p)=\Gamma_{1,1}^c\cos\tau+\Gamma_{1,1}^s\sin\tau+\Gamma_{3,1}^c\cos3\tau+\Gamma_{3,1}^s\sin3\tau$$
$$+\sum_{i=2}p^{i-1}\{\Gamma_{2i+1,1}^c\cos[(2i+1)\tau]+\Gamma_{2i+1,1}^s\sin[(2i+1)\tau]\}\tag{1.12}$$

We separate the Fourier series in Eq. (1.12) into two parts: the first part is the main residues considered to find the first approximation and the second part under the summation sign is the remaining residues to find the next approximation. The right hand side of Eq. (1.12) should not contain the terms $\sin(\tau)$, $\cos(\tau)$, $\sin(3\tau)$ and $\cos(3\tau)$ based on Galerkin technique. Letting their coefficients be zeros, yields

$$\Gamma_{j,1}^c(a_{1,1},a_{3,1},b_{3,1},\omega_1)=\frac{1}{\pi}\int_0^{2\pi}R_u^1(\tau)\cos(j\tau)\mathrm{d}\tau=0$$
$$\Gamma_{j,1}^s(a_{1,1},a_{3,1},b_{3,1},\omega_1)=\frac{1}{\pi}\int_0^{2\pi}R_u^1(\tau)\sin(j\tau)\mathrm{d}\tau=0\tag{1.13}$$
$$j=1,3.$$

Thus, the linear Eq. (1.13) includes four equations and four unknowns that can be solved easily. It should be pointed that the coefficients of sine function and $b_{3,1}$ become zero when Eq. (1.1) is a conservative system and $a_{1,1}+a_{3,1}=0$.

The second-order analytical approximation ($m=2$) can be obtained by substituting Eqs. (1.6), (1.7), (1.9) and (1.12) into Eq. (1.3) and equating the coefficients of the p^2,

$$\left(\omega_2\frac{\partial}{\partial\omega}+u_2''\frac{\partial}{\partial u''}+u_2'\frac{\partial}{\partial u'}+u_2\frac{\partial}{\partial u}\right)\Phi_0+\frac{1}{2!}\left(\omega_1\frac{\partial}{\partial\omega}+u_1''\frac{\partial}{\partial u''}+u_1'\frac{\partial}{\partial u'}+u_1\frac{\partial}{\partial u}\right)^2\Phi_0$$

$$+(\Gamma_{5,0}^c+\Gamma_{5,1}^c)\cos(5\tau)+(\Gamma_{5,0}^s+\Gamma_{5,1}^s)\sin(5\tau)=R_u^2(\tau) \tag{1.14}$$

Then rewrite the residues $R_u^2(\tau)$ to $R_u^2(\tau,p)$,

$$R_u^2(\tau,p)=\Gamma_{1,2}^c\cos\tau+\Gamma_{1,2}^s\sin\tau+\Gamma_{3,2}^c\cos 3\tau+\Gamma_{3,2}^s\sin 3\tau+\Gamma_{5,2}^c\cos 5\tau+\Gamma_{5,2}^s\sin 5\tau$$

$$+\sum_{i=3}p^{i-2}\{\Gamma_{2i+1,2}^c\cos[(2i+1)\tau]+\Gamma_{2i+1,2}^s\sin[(2i+1)\tau]\}$$

$$\tag{1.15}$$

We separate the Fourier series in Eq. (1.15) into two parts: the first part is the main residues considered to find the second approximation and the second part under the summation sign is the remaining residues to find the third or higher approximation. The right hand side of Eq. (1.15) should not contain the terms $\sin(\tau)$, $\cos(\tau)$, $\sin(3\tau)$, $\cos(3\tau)$, $\sin(5\tau)$ and $\cos(5\tau)$ based on Galerkin technique. Letting their coefficients be zeros, yields

$$\Gamma_{j,2}^c(a_{1,2},a_{3,2},b_{3,2},a_{5,2},b_{5,2},\omega_2)=\frac{1}{\pi}\int_0^{2\pi}R_u^2(\tau)\cos(j\tau)\mathrm{d}\tau=0, j=1,3,5,$$

$$\Gamma_{j,2}^s(a_{1,2},a_{3,2},b_{3,2},a_{5,2},b_{5,2},\omega_2)=\frac{1}{\pi}\int_0^{2\pi}R_u^2(\tau)\sin(j\tau)\mathrm{d}\tau=0, j=1,3,5. \tag{1.16}$$

The linear Eq. (1.16) which including six equations and six unknowns can be easily solved by means of the computational software such as Maple or MatLab algorithm. Similar to analysis above mentioned, the coefficients of sine function and $b_{3,2}$, $b_{5,2}$ become zero when Eq. (1.1) is a conservative system and $a_{1,2}+a_{3,2}+a_{5,2}=0$.

In general, one defines the operator

$$D_k=\omega_k\frac{\partial}{\partial\omega}+u_k''\frac{\partial}{\partial u''}+u_k'\frac{\partial}{\partial u'}+u_k\frac{\partial}{\partial u}, \text{ So that}$$

$$D_k^n=\left(\omega_k\frac{\partial}{\partial\omega}+u_k''\frac{\partial}{\partial u''}+u_k'\frac{\partial}{\partial u'}+u_k\frac{\partial}{\partial u}\right)^n$$

The residue $R_u^j(\tau)$ in the j-th step is defined by

$$\left(D_j + \frac{1}{2!}D_{j-1}^2 + \cdots + \frac{1}{j!}D_1^j\right)\Phi_0 + \sum_{i=0}^{j-1}\{\Gamma_{2j+1,i}^c \cos\left[(2j+1)\tau\right] + \Gamma_{2j+1,i}^s \sin\left[(2j+1)\tau\right]\} = R_u^j(\tau)$$

These are linear equations with respect to ω_j and u_j. The Fourier expansion of $R_u^j(\tau)$ is

$$R_u^j(\tau,p) = \sum_{i=0}^{j}\Gamma_{2i+1,j}^c \cos[(2i+1)\tau] + \Gamma_{2i+1,j}^s \sin[(2i+1)\tau]$$

$$+ \sum_{i=j+1} p^{i-j}\{\Gamma_{2i+1,j}^c \cos[(2i+1)\tau] + \Gamma_{2i+1,j}^s \sin[(2i+1)\tau]\}$$

To eliminate the terms corresponding to each term of the first summation, there are the same number of linear equations for the same number of unknowns $a_{i,j}, b_{i,j}, \omega_j$, $i = 1,3,5,\cdots$, that one can obtain the solutions to any desired accuracy.

Remark: For the conservative oscillator with even nonlinearities, we may transform the equation under study into two parts in positive and negative directions by introducing a sign function. The initial amplitude in negative direction can be obtained from energy conservative. Then we use symmetry form to express the solution of equation with even nonlinearity.

1.2.2 Non-autonomous Differential Equation

Consider a strongly nonlinear non-autonomous system of m DOF (degrees-of-freedom) with periodic excitation and response in the form

$$\Phi[\ddot{u},\dot{u},u,\lambda,f(\omega t)] = 0 \tag{1.17}$$

where a dot denotes differentiation with respect to t, λ is a system parameter, ω is the frequency of excitation and u is a vector of m unknowns. For convenience, we introduce a new time scale $\tau = \omega t$,

$$\Phi[\omega^2 u'',\omega u',u,\lambda,f(\tau)] = 0 \tag{1.18}$$

where prime denotes differentiation with respect to τ.

Periodic solutions of differential equations can be well approximated by Fourier series [Mickens, 1996]. Thus, the harmonic balance solution $u_0(\tau)$ using n harmonic terms can be described as

$$u_0(\tau) = \sum_{i=0}^{n} a_{i,0} \cos(i\tau) + \sum_{i=1}^{n} b_{i,0} \sin(i\tau) \tag{1.19}$$

Substituting Eq. (1.19) into Eq. (1.18) and using total harmonic balance, one has

$$\Phi[\omega^2 u_0'', \omega u_0', u_0, \lambda, f(\tau)] = \sum_{i=0}^{\kappa n} R_{i,0}^{c} \cos(i\tau) + \sum_{i=1}^{\kappa n} R_{i,0}^{s} \sin(i\tau)$$

where κ is an integer depending on the nature of the nonlinearity. For linear problem, $\kappa = 1$. When the nonlinearity is polynomial with the highest power term $\ddot{u}^{\kappa_1}\dot{u}^{\kappa_2}u^{\kappa_3}$, $\kappa = \kappa_1 + \kappa_2 + \kappa_3$, for integers κ_1, κ_2 and κ_3. For rational or irrational powers, $\ddot{u}^{\kappa_1}\dot{u}^{\kappa_2}u^{\kappa_3}$ will be expanded in Fourier series of infinite number of terms, κ is infinity to be exact. Without loss of generality, one shall use $\kappa = 2$ for convenience of presentation in the following section. Therefore,

$$\Phi[\omega^2 u_0'', \omega u_0', u_0, \lambda, f(\tau)] = \sum_{i=0}^{n} R_{i,0}^{c} \cos(i\tau) + \sum_{i=1}^{n} R_{i,0}^{s} \sin(i\tau) + \sum_{i=n+1}^{2n} R_{i,0}^{c} \cos(i\tau)$$

$$+ \sum_{i=n+1}^{2n} R_{i,0}^{s} \sin(i\tau)$$

One has the following $m(2n+1)$ harmonic balance equations for the residues R_i^{c} and R_i^{s},

$$R_i^{c}(a_{0,0}, a_{1,0}, \cdots, a_{n,0}; b_{1,0}, \cdots, b_{n,0}) = \frac{2}{\pi} \int_0^{\pi} \Phi[\omega^2 u_0'', \omega u_0', u_0, \lambda, f(\tau)] \cos(i\tau) \mathrm{d}\tau = 0, \ i = 0, \cdots, n$$

$$R_i^{s}(a_{0,0}, a_{1,0}, \cdots, a_{n,0}; b_{1,0}, \cdots, b_{n,0}) = \frac{2}{\pi} \int_0^{\pi} \Phi[\omega^2 u_0'', \omega u_0', u_0, \lambda, f(\tau)] \sin(i\tau) \mathrm{d}\tau = 0, \ i = 1, \cdots, n$$

$$\tag{1.20}$$

The unknown coefficients $a_{0,0}, a_{1,0}, \cdots, a_{n,0}, b_{1,0}, \cdots, b_{n,0}$ can be obtained by solving the above nonlinear algebraic equations. The second subscript 0 after comma denotes initial solutions using n harmonics truncated to $m(2n+1)$ equations.

When substituting the obtained solutions (1.19) into the Eq. (1.18), the higher residues for $i > n$ are not necessarily equal to zero due to the neglected higher harmonics. That is,

$$R_i^c(a_{0,0}, a_{1,0}, \cdots, a_{n,0}, b_{1,0}, \cdots, b_{n,0}) = \frac{2}{\pi} \int_0^\pi \Phi[\omega^2 u_0'', \omega u_0', u_0, \lambda, f(\tau)] \cos(i\tau) \mathrm{d}\tau \neq 0, \ i > n$$

$$R_i^s(a_{0,0}, a_{1,0}, \cdots, a_{n0}, b_{1,0}, \cdots, b_{n,0}) = \frac{2}{\pi} \int_0^\pi \Phi[\omega^2 u_0'', \omega u_0', u_0, \lambda, f(\tau)] \sin(i\tau) \mathrm{d}\tau \neq 0, \ i > n$$

$$\tag{1.21}$$

If $u(\tau)$ is an assumed initial approximate solution of Eq.(1.18), then the residue when considering the neglected higher harmonics can be written as

$$\tilde{R}_0(\tau) = \Phi[\omega^2 u_0'', \omega u_0', u_0, \lambda, f(\tau)] \neq 0$$

$$\Phi[\omega^2 u_0'', \omega u_0', u_0, \lambda, f(\tau)] = \sum_{i=0}^n R_i^c \cos i\tau + \sum_{i=1}^n R_i^s \sin i\tau$$

$$+ p\left(\sum_{i=n+1}^{2n} R_i^c \cos i\tau + \sum_{i=n+1}^{2n} R_i^s \sin i\tau \right)$$

$$= p\tilde{R}_0$$

$$\tag{1.22}$$

where p is an order parameter which may be taken as zero when the residue is not considered and as one for retaining all residues. Applying the harmonic balance for $p = 0$, one has $m(2n + 1)$ nonlinear algebraic equations for the determination of the initial approximation of the same number of unknowns $a_{0,0}, a_{1,0}, \cdots, a_{n,0}, b_{1,0}, \cdots, b_{n,0}$. It would be rather difficult to do so when n is larger. Moreover, for motions with evident higher harmonic components, more harmonics have to be included in the analysis of the assumed initial approximation. When harmonic terms n is larger, the solution procedure of harmonic balance equations (1.20) can be found in section 1.3 by introducing polynomial homotopy continuation technique to obtain all impossible solutions, so that bifurcation analysis can be implemented. Once the assumed initial solution (1.19) is solved, the procedure of residue harmonic balance can be progressed similarly to autonomous case to improve the accuracy of the steady state responses. The following is devoted to amending the assumed steady state responses.

For the assumed steady state response u_0, writing

$$\Phi[\omega^2 u_0'', \omega u_0', u_0, \lambda, f(\tau)] = \sum_{i=0}^n R_i^c \cos i\tau + \sum_{i=1}^n R_i^s \sin i\tau + p\tilde{R}_0$$

Then, the forward residue $\tilde{R}_0$ is useful for the determination of the first approximation below.

To find the first approximation $u_{(1)}(\tau) = u_0(\tau) + u_1(\tau)$, an order parameter p having values in the interval is introduced with an aim to improve the solution from $u_0(\tau)$ to $u_1(\tau)$ where $u_1(\tau)$ be the first correction to be determined. The embedded forms of variables $u(\tau)$ and $\tilde{R}_0(\tau)$ can be written as $u_{(1)}(\tau, p)$ and $\tilde{R}_0(\tau, p)$ respectively resulting in

$$u_{(1)}(\tau, p) = u_0(\tau) + pu_1(\tau), \quad \tilde{R}_0(\tau, p) = p\tilde{R}_0(\tau) \tag{1.23}$$

$u_1(\tau)$ has the same period as $u_0(\tau)$, then, its Fourier representation is

$$u_1(\tau) = \sum_{i=0}^{2n} a_{i,1} \cos i\tau + \sum_{i=1}^{2n} b_{i,1} \sin i\tau \tag{1.25}$$

and

$$u_{(1)}(\tau) = u_0(\tau) + u_1(\tau) = \sum_{i=0}^{n} a_{i,0} \cos(i\tau) + \sum_{i=1}^{n} b_{i,0} \sin(i\tau) + \sum_{i=0}^{2n} a_{i,1} \cos(i\tau) + \sum_{i=1}^{2n} b_{i,1} \sin(i\tau)$$

where $a_{0,1}, a_{i,1}, b_{i,1}, i = 1, \cdots, 2n$ are the unknown Fourier coefficients of $u_1(\tau)$. Substituting Eq. (1.23) into Eq. (1.17) and equating similar terms in p, one has a set of linear ODE to determine $u_1(\tau)$,

$$\left(u_1'' \frac{\partial}{\partial u''} + u_1' \frac{\partial}{\partial u'} + u_1 \frac{\partial}{\partial u} \right) \Phi \, |_{u=u_0} + \tilde{R}_0(\tau) = 0 \tag{1.25}$$

One rewrites the equation to $\tilde{R}_0(\tau) + \left(\sum_{k=0}^{2} u_1^{(k)} \frac{\partial}{\partial u^{(k)}} \right) \Phi \, |_0 = 0$

for the convenience in later presentation, where superscript in brackets denotes the order of derivatives with respect to τ and $|_0$ denotes that $\dfrac{\partial \Phi}{\partial u^{(k)}}$ is to be evaluated at $u=u_0$ after differentiation. Substituting Eq. (1.24) into the linear Eq. (1.25) and using harmonic balance by multiplying Eq. (1.25) by $\cos(i\tau)$ for $i=0$ to $2n$ and $\sin(i\tau)$ for $i=1$ to $2n$ and integrating over one period, one has $m(4n+1)$ linear equations for the unknown Fourier coefficients $a_{i,1}, b_{i,1}$ in Eq. (1.24). According to Eqs. (1.19), (1.23), (1.24) and letting $p=1$, the first-level improved solution can be obtained as

$$u_{(1)}(\tau) = u_0(\tau) + u_1(\tau) = \sum_{i=0}^{n}(a_{i,0} + a_{i,1})\cos i\tau + \sum_{i=1}^{n}(b_i + b_{i,1})\cos i\tau + \sum_{i=n+1}^{2n}(a_{i,1}\cos i\tau + b_{i,1}\sin i\tau)$$

$$(1.26)$$

Similarly, substituting the first-level improved solutions into the Eq. (1.18), a new residue $\tilde{R}_1(\tau)$ will be created. In seeking the more accurate second approximation, one writes

$$u_{(2)}(\tau, p) = u_0(\tau) + pu_1(\tau) + p^2 u_2(\tau), \quad \tilde{R}_1(\tau, p) = p^2 \tilde{R}_1(\tau)$$

$$(1.27)$$

Substituting Eq. (1.27) into Eq. (1.17) and equating terms of p^2 gives

$$\tilde{R}_1(\tau) + \left[\sum_{k=0}^{2}u_2^{(k)}\frac{\partial}{\partial u^{(k)}} + \frac{1}{2!}\left(\sum_{k=0}^{2}u_1^{(k)}\frac{\partial}{\partial u^{(k)}}\right)^2\right]\Phi\,|_0 = 0$$

$$(1.28)$$

where, denoting $u_1^{(k)}$ by $v^{(k)}$ for convenience, the square of the vector inside the brackets is defined by $\left(\sum_{k=0}^{2}v^{(k)}\dfrac{d}{du^{(k)}}\right)^2\Phi = \left(\sum_{k=0}^{2}\sum_{l=0}^{2}v^{(k)}v^{(l)}\dfrac{\partial^2}{\partial u^{(k)}\partial u^{(l)}}\right)\Phi$. An explicit formulae for $m=2$ as reference is given,

$$\sum_{k=0}^{2}\sum_{l=0}^{2}v^{(k)}v^{(l)}\frac{\partial^2\Phi}{\partial u^{(k)}\partial u^{(l)}} = \sum_{k=0}^{2}\sum_{l=0}^{2}\sum_{i=1}^{m}\sum_{j=1}^{m}v_i^{(k)}v_j^{(l)}\frac{\partial^2\Phi}{\partial u_i^{(k)}\partial u_j^{(l)}} = (\frac{\partial\Phi}{\partial u_1''}v_1'' + \frac{\partial\Phi}{\partial u_2''}v_2'' + \frac{\partial\Phi}{\partial u_1'}v_1'$$

$$+\frac{\partial\Phi}{\partial u_2'}v_2' + \frac{\partial\Phi}{\partial u_1}v_1 + \frac{\partial\Phi}{\partial u_2}v_2)^2$$

$$=\frac{\partial^2\Phi}{\partial u_1''^2}v_1''^2 + \frac{\partial^2\Phi}{\partial u_2''^2}v_2''^2 + \frac{\partial^2\Phi}{\partial u_1'^2}v_1'^2 + \frac{\partial^2\Phi}{\partial u_2'^2}v_2'^2 + \frac{\partial^2\Phi}{\partial u_1^2}v_1^2 + \frac{\partial^2\Phi}{\partial u_2^2}v_2^2 + 2(\frac{\partial^2\Phi}{\partial u_1''\partial u_2''}v_1''v_2''$$

$$+\frac{\partial^2\Phi}{\partial u_1''\partial u_1'}v_1''v_1' + \frac{\partial^2\Phi}{\partial u_1''\partial u_2'}v_1''v_2' + \frac{\partial^2\Phi}{\partial u_1''\partial u_1}v_1''v_1 + \frac{\partial^2\Phi}{\partial u_1''\partial u_2}v_1''v_2 + \frac{\partial^2\Phi}{\partial u_2''\partial u_1'}v_2''v_1' + \frac{\partial^2\Phi}{\partial u_2''\partial u_2'}v_2''v_2'$$

$$+\frac{\partial^2\Phi}{\partial u_2''\partial u_1}v_2''v_1 + \frac{\partial^2\Phi}{\partial u_2''\partial u_2}v_2''v_2 + \frac{\partial^2\Phi}{\partial u_1'\partial u_2'}v_1'v_2' + \frac{\partial^2\Phi}{\partial u_1'\partial u_1}v_1'v_1 + \frac{\partial^2\Phi}{\partial u_1'\partial u_2}v_1'v_2 + \frac{\partial^2\Phi}{\partial u_2'\partial u_1}v_2'v_1$$

$$+\frac{\partial^2\Phi}{\partial u_2'\partial u_2}v_2'v_2 + \frac{\partial^2\Phi}{\partial u_1\partial u_2}v_1v_2)$$

$u_2(\tau)$ has the same period as $u_0(\tau)$, then its Fourier expansion is

$$u_2(\tau) = \sum_{i=0}^{3n}a_{i,2}\cos i\tau + \sum_{i=1}^{3n}b_{i,2}\sin i\tau$$

$$(1.29)$$

$$\tilde{R}_1(\tau)+\left[\sum_{k=0}^{2}u_2^{(k)}\frac{\partial}{\partial u^{(k)}}+\frac{1}{2!}\left(\sum_{k=0}^{2}u_1^{(k)}\frac{\partial}{\partial u^{(k)}}\right)^2\right]\Phi\,|_0=\sum_{i=0}^{4n}R_{i,2}^c\cos i\tau+\sum_{i=1}^{4n}R_{i,2}^s\sin i\tau$$

$$=\sum_{i=0}^{3n}R_{i,2}^c\cos i\tau+\sum_{i=1}^{3n}R_{i,2}^s\sin i\tau$$

$$+p\tilde{R}_2$$

where $a_{0,2},a_{i,2},b_{i,2},i=1,\cdots,3n$ are the unknown Fourier coefficients, substituting Eq. (1.29) into the linear Eq. (1.28) and using harmonic balance method, the unknown Fourier coefficients $a_{i,2},b_{i,2}$ can be easily obtained by solving $m(6n+1)$ linear algebraic equations. The forward residue $\tilde{R}_2$ is useful for the determination of the third approximation.

According to Eqs. (1.19), (1.23), (1.25), (1.29) and writing $p=1$, the second approximation can be obtained as follows .

$$u_{(2)}(\tau)=u_0(\tau)+u_1(\tau)+u_2(\tau)\tag{1.30}$$

By mathematical induction, one can prove that when the $(r-1)$-th approximation is available, the r-th approximation is given by

$$u_{(r)}(\tau)=u_0(\tau)+\cdots+u_r(\tau)\tag{1.31}$$

Let

$$u_r(\tau)=\sum_{i=0}^{(r+1)n}a_{i,r}\cos i\tau+\sum_{i=1}^{(r+1)n}b_{i,r}\sin i\tau\tag{1.32}$$

$$\tilde{R}_{r-1}(\tau)+\left[\sum_{k=0}^{2}u_r^{(k)}\frac{\partial}{\partial u^{(k)}}+\frac{1}{2!}\left(\sum_{k=0}^{2}u_{r-1}^{(k)}\frac{\partial}{\partial u^{(k)}}\right)^2+\cdots+\frac{1}{r!}\left(\sum_{k=0}^{2}u_1^{(k)}\frac{\partial}{\partial u^{(k)}}\right)^r\right]\Phi\,|_0$$

$$=\sum_{i=0}^{(r+2)n}R_{i,r}^c\cos i\tau+\sum_{i=1}^{(r+2)n}R_{i,r}^s\sin i\tau=\sum_{i=0}^{(r+1)n}R_{i,r}^c\cos i\tau+\sum_{i=1}^{(r+1)n}R_{i,r}^s\sin i\tau+p\tilde{R}_r\tag{1.33}$$

where $a_{0,r},a_{i,r},b_{i,r},i=1,\cdots,(r+1)n$ are the unknown Fourier coefficients, substituting Eq. (1.32) into the linear Eq. (1.33) when $p=0$ and using harmonic balance method, the unknown Fourier coefficients $a_{i,2},b_{i,2}$ can be easily obtained by solving $m[2(r+1)n+1]$ linear algebraic equations. The forward residue $\tilde{R}_r$ is useful for the determination of the $(r+1)$ th approximation.

1.2.3 Error and Convergence Analysis

As a variant of classical harmonic balance method, the convergence analysis for classical harmonic balance method was studied and proved by M. Urabe (1965); M. Urabe & A. Reiter (1966); Kelley &Mukundan (1993). According to the Banach's fixed-point theorem, the following conclusions can be proved.

Theorem 1.1: The series solution with $p = 1$ for residue harmonic balance approximations,

$$u(t) = \sum_{i=0}^{\infty} u_i(t) \tag{1.34}$$

converges if there is $0 < \eta < 1$ such that $\left\| u_{i+1}(t) \right\| \leq \eta \left\| u_i(t) \right\|$ for $\forall i \geq i_0$, and $i_0 \in N$.

Proof. We first define the sequence $\{s_n\}$ as

$$s_n = \sum_{i=0}^{n} u_i(t), n = 0,1,2,\cdots \tag{1.35}$$

Then

$$\left\| s_{n+1} - s_n \right\| = \left\| u_{n+1} \right\| \leq \eta \left\| u_n \right\| \leq \cdots \leq \eta^{n-i_0+1} \left\| u_{i_0} \right\|$$

And, for every $j, k \in N, j \geq k > i_0$

$$\left\| s_j - s_k \right\| = \left\| (s_j - s_{j-1}) + (s_{j-1} - s_{j-2}) + \cdots + (s_{k+1} - s_k) \right\|$$

$$\leq \left\| s_j - s_{j-1} \right\| + \left\| s_{j-1} - s_{j-2} \right\| + \cdots + \left\| s_{k+1} - s_k \right\|$$

$$= \left\| u_j \right\| + \left\| u_{j-1} \right\| + \cdots + \left\| u_{k+1} \right\|$$

$$\leq \eta^{j-i_0} \left\| u_{i_0} \right\| + \eta^{j-1-i_0} \left\| u_{i_0} \right\| + \cdots + \eta^{k+1-i_0} \left\| u_{i_0} \right\|$$

$$= \frac{1-\eta^{j-k}}{1-\eta} \eta^{k+1-i_0} \left\| u_{i_0} \right\|$$

Consequently

$$0 \leq \lim_{j,k \to \infty} \left\| s_j - s_k \right\| \leq \lim_{j,k \to \infty} \frac{1-\eta^{j-k}}{1-\eta} \eta^{k+1-i_0} \left\| u_{i_0} \right\| = 0 \tag{1.36}$$

Since $0 < \eta < 1$, we have

$$\lim_{j,k \to \infty} \left\| s_j - s_k \right\| = 0$$

From the definition of Cauchy sequence in the Hilbert space, sequence $\{s_n\}$ is a Cauchy sequence that implies the series solution $\sum_{i=0}^{\infty} u_i(t)$ converges.

Theorem 1.2: If the series solution $\sum_{i=0}^{\infty} u_i(t)$ is convergent to the solution $u_i(t)$ and we take the mth-order truncated series $\sum_{i=0}^{\infty} u_i(t)$ as an approximation to the solution $u_i(t)$, then the maximum absolute truncated error is estimated as

$$\left\| u(t) - \sum_{i=0}^{m} u_i(t) \right\| \leq \frac{1}{1-\eta} \eta^{m+1} \left\| u_0(t) \right\| \tag{1.37}$$

1.3 Polynomial Homotopy Continuation Technique

The harmonic balance method truncates the Fourier series in a finite number of terms. For some motions, we need to consider superharmonic and subharmonic resonance responses of system and it is necessary to keep all nonlinear terms so that bifurcation analysis can be implemented. Then we are encountered with the multiple solutions complexity of the nonlinear harmonic balance equations. For example, a cubic equation has three roots in general and eight cubic equations will have at most $3^8 = 6561$ solutions. To obtain all solutions together with illustrating the bifurcation of all real solutions is a difficult task. Therefore, many researchers used Newton method to solve these harmonic balance equations. The traditional Newton method is a well-known and popular method for solving nonlinear equations. It has high efficient in the convergence speed. However, we always need to guess the initial value in the iteration process. Good initial guess value can solve the equation quickly. Bad initial guess value usually will yield divergence. Besides, due to the arbitrariness in the selection of initial conditions, some solutions may be lost.

Homotopy continuation method places a higher priority on finding possibly all solutions reliably than on finding one or a few solutions. The idea is to break the harmonic balance equation into a simpler starting problem that has the same number of solutions by constructing a start system embedded in a homotopy. Then, one can trace the solution paths defined by the homotopy using predictor-corrector method. The following reviews the main idea of polynomial homotopy continuation method by a simple example.

Consider a simultaneous nonlinear algebraic equation of form

$$F(U) = 0, U = (u_1, u_2, \cdots, u_n)^T \tag{1.38}$$

To solve Eq. (1.38), one chooses a new simple start system or call auxiliary homotopy function

$$G(U) = 0, U = (u_1, u_2, \cdots, u_n)^T \tag{1.39}$$

The auxiliary homotopy function $G(U)$ is known or controllable and easy to solve. Then, one defines the homotopy continuation function as

$$H(U, p) = pF(U) + (1-p)G(U) = 0, U = (u_1, u_2, \cdots, u_n)^T \tag{1.40}$$

where p is an arbitrary parameter and varies from 0 to 1, i.e. $p \in [0,1]$. Therefore, one has the following two boundary conditions

$$H(U,0) = G(U); H(U,1) = F(U) \tag{1.41}$$

Then our goal is to solve the $H(U,p) = 0$ instead of $F(U) = 0$ by varying parameter p from 0 to 1. To avoid divergence, Wu [T.M. Wu, 2005] provided some useful choices about the auxiliary homotopy function. By appropriate adjusting the auxiliary homotopy function, the solution of Eq. (1.38) can be obtained. In this thesis, we mainly adopt the procedure from Li [T.Y. Li, 1997] and Sommese [A.J. Sommese et al., 2005a, 2005b].

Chapter 2 Asymptotic Analysis of Autonomous Fractional Ordinary Differential Systems

2.1 Oscillatory Region and Asymptotic Solution of Fractional Van der Pol Oscillator

2.1.1 Introduction

Since the original work of Balthazar van der Pol in 1920s [B. van der Pol, 1920], the van der Pol oscillator, served as a basic model of self-sustained oscillations, has been widely used in the mechanical, electronic engineering, biological sciences and so on. Recently, Pereira et al [E. Pereira et al., 2004] substituted the capacitance by a fractance in the nonlinear RLC circuit to obtain a fractional version of the van der Pol equation. Barbosa et al [R.S.M. Barbosa et al., 2004, 2007] introduced a more general modified version of fractional order van der Pol oscillator. In recent years, the study of oscillatory behaviors in fractional order dynamical systems has been a subject of increasing attention [R.S.M. Barbosa et al., 2004, 2007; I. Podlubny, 1999; L. Debnath, 2003; W. Ahmed et al., 2001; Y. Wang & C. Li, 2007; S. Chen et al., 2009; Y.A. Rossikhin & M.V. Shitikova, 2010]. It has been found that the fractional system like integer order systems can generate regular oscillations. For instance, Ahmad et al [W. Ahmed et al., 2001] pointed out that limit cycle could be generated in the fractional Wien bridge oscillator. Wang and Li [Y. Wang & C. Li, 2007] numerically showed the existence of limit cycle for fractional Brusselator. Thus, the response and dynamical behavior of the fractional differential systems has started to attract attention in the last few decades. Owing to the presence of nonlinearities and fractional derivative in fractional van der Pol, explicit solutions are rarely obtained.

The aim of this section to predict the parametric function for the boundary between oscillatory and non-oscillatory regions of the fractional van der Pol oscillator by the first-order approximation using just one Fourier term, and then to obtain the highly accurate solutions for steady state of fractional van der Pol oscillator by extending residue harmonic balance method. In addition, the effects of the fractional order and control parameter on the angular frequency and amplitude of limit cycle are illustrated and discussed.

2.1.2 Fractional Van der Pol Oscillator

The van der Pol oscillator is given by a self-excited differential equation in its standard form

$$\ddot{u}(t) + \varepsilon[u(t)^2 - 1]\dot{u}(t) + u(t) = 0 \tag{2.1}$$

where ε is a control parameter that reflects the degree of nonlinear of the system and it possesses a periodic solution in steady state.

Recently, by substituting the capacitance by a fractance in the nonlinear RLC circuit, Pereira et al. [E. Pereira et al., 2004] obtained a fractional version of the van der Pol equation

$$D_t^\lambda u(t) + \varepsilon[u(t)^2 - 1]\dot{u}(t) + u(t) = 0 \tag{2.2}$$

where $1 < \lambda < 2$.

More recently, Xie and Lin [F. Xie & X.Y. Lin, 2009] introduced the following van der Pol oscillator with a small fractional damping term and studied the asymptotic solution of Cauchy problem.

$$\ddot{u}(t) + \varepsilon[u(t)^2 - 1]D_t^\lambda u(t) + u(t) = 0, 0 < \varepsilon \ll 1$$

$$u(0) = a, \dot{u}(0) = b \tag{2.3}$$

where $0 < \lambda < 1$.

In this section, a more general modified version of van der Pol oscillator (R.S. M. Barbosa et al., 2004, 2007)

$$D_t^{1+\lambda}u(t) + \varepsilon[u(t)^2 - 1]D_t^{\lambda}u(t) + u(t) = 0 \tag{2.4}$$

is considered, where u is displacement, ε is a control parameter that reflects the degree of nonlinear of the system and $\varepsilon > 0$, $0 < \lambda < 1$ is the order of the fractional derivative. $D_t^{1+\lambda}$ and D_t^{λ} are the fractional derivative in the Caputo sense is defined by

$$D_t^{\alpha}u(t) = \frac{1}{\Gamma(m-\alpha)} \int_0^t (t-\tau)^{m-\alpha-1} \frac{d^m}{d\tau^m} u(\tau) d\tau \tag{2.5}$$

for $m-1 < \alpha \le m, m \in N, t > 0, u \in C_{-1}^m$.

Barbosa showed the steady state time responses and Fourier spectra of Eq. (2.4) through numerical simulation. Tavazoei et al [M.S. Tavazoei et al., 2009] determined parametric range of Eq. (2.4) as an undamped oscillator and analyzed its more details.

2.1.3 Analysis of Oscillatory Parametric Regions

2.1.3.1 Steady state response

Due to the symmetric of Eq. (2.4), we assume that the angular frequency of Eq. (2.4) is ω and the steady state response of form

$$u_0(t) = a_1 \cos(\omega t) + \sum_{i=1}^{n} \{a_{2i+1} \cos[(2i+1)\omega t] + b_{2i+1} \sin[(2i+1)\omega t\} \tag{2.6}$$

A new independent variable is $\tau = \omega t$ introduced, so that the new period is known to be 2π, then Eqs. (2.4) and (2.6) can be rewritten as follows:

$$\omega^{1+\lambda}D_t^{1+\lambda}u(\tau) + \varepsilon\omega^{\lambda}[u(\tau)^2 - 1]D_{\tau}^{\lambda}u(\tau) + u(\tau) = 0 \tag{2.7}$$

$$u_0(\tau) = a_1 \cos(\tau) + \sum_{i=1}^{n} \{a_{2i+1} \cos[(2i+1)\tau] + b_{2i+1} \cos[(2i+1)\tau]\} \tag{2.8}$$

Based on definition (2.5), the fractional derivatives $D_\tau^\beta u(\tau), (0 < \beta < 1)$ can be computed as below

$$D_\tau^\beta u(\tau) \overset{\Delta}{=} D_\tau^\beta [a_k \cos(k\tau) + b_k \sin(k\tau)] = \frac{1}{\Gamma(1-\beta)} \int_0^\tau (\tau - s)^{-\beta} \frac{\mathrm{d}}{\mathrm{d}s} [a_k \cos(ks) + b_k \sin(ks)] \mathrm{d}s$$

$$= \frac{k}{\Gamma(1-\beta)} \int_0^\tau t^{-\beta} [b_k \cos(k\tau - kt) - a_k \sin(k\tau - kt)] \mathrm{d}t$$

$$(2.9)$$

According to the following trigonometric identities:

$$\cos(k\tau - kt) = \cos k\tau \cos kt + \sin k\tau \sin kt$$

and $\sin(k\tau - kt) = \sin k\tau \cos kt - \cos k\tau \sin kt$.

Eq. (2.9) becomes

$$D_\tau^\beta u(\tau) \overset{\Delta}{=} \frac{k^\beta}{\Gamma(1-\beta)} [\cos(k\tau) \int_0^{k\tau} \upsilon^{-\beta} (b_k \cos\upsilon + a_k \sin\upsilon) \mathrm{d}\upsilon + \sin(k\tau) \int_0^{k\tau} \upsilon^{-\beta} (b_k \sin\upsilon - a_k \cos\upsilon) \mathrm{d}\upsilon]$$

$$= k^\beta [b_k \cos(k\tau) - a_k \sin(k\tau)] \int_0^{k\tau} \frac{\upsilon^{-\beta} \cos\upsilon}{\Gamma(1-\beta)} \mathrm{d}\upsilon + k^\beta [a_k \cos(k\tau) + b_k \sin(k\tau)] \int_0^{k\tau} \frac{\upsilon^{-\beta} \sin\upsilon}{\Gamma(1-\beta)} \mathrm{d}\upsilon$$

$$= k^\beta (b_k I_1 + a_k I_2) \cos(k\tau) + k^\beta (b_k I_2 - a_k I_1) \sin(k\tau)$$

$$(2.10)$$

where

$$I_1 = \int_0^{k\tau} \frac{\upsilon^{-\beta} \cos\upsilon}{\Gamma(1-\beta)} \mathrm{d}\upsilon, I_2 = \int_0^{k\tau} \frac{\upsilon^{-\beta} \sin\upsilon}{\Gamma(1-\beta)} \mathrm{d}\upsilon .$$

Although these integrals cannot be evaluated in closed form for general interval, we are able to evaluate them using the contour integration, due to the research of steady state response, where

$$I_1^\infty = \lim_{\tau \to \infty} I_1 = \frac{\sqrt{\pi} \Gamma[(1-\beta)/2]}{2^\beta \Gamma(\beta/2) \Gamma(1-\beta)} = \sin\frac{\beta\pi}{2},$$

$$I_2^\infty = \lim_{\tau \to \infty} I_2 = \frac{\sqrt{\pi} \Gamma(1-\beta/2)}{2^\beta \Gamma[(1+\beta)/2] \Gamma(1-\beta)} = \cos\frac{\beta\pi}{2}.$$

In the following calculation, we substitute the I_1 I_2 in Eq. (2.10) by I_1^∞, I_2^∞ respectively.

Substituting Eq. (2.8) into Eq. (2.7) and using the Galerkin procedure result in the following harmonic balance equations.

$$R_i^c(\omega,a_1,a_3,\cdots,a_{2n+1},b_3,\cdots,b_{2n+1}) = \frac{2}{\pi}\int_0^\pi [\omega^{1+\lambda}D_\tau^{1+\lambda}u + a\omega^\lambda(u^2-1)D_\tau^\lambda u + u]\cos(2i+1)\tau\,\mathrm{d}\tau$$

$$= 0, i = 0,1,\cdots,n$$

$$R_i^s(\omega,a_1,a_3,\cdots,a_{2n+1},b_3,\cdots,b_{2n+1}) = \frac{2}{\pi}\int_0^\pi [\omega^{1+\lambda}D_\tau^{1+\lambda}u + a\omega^\lambda(u^2-1)D_\tau^\lambda u + u]\sin(2i+1)\tau\,\mathrm{d}\tau$$

$$= 0, i = 0,1,\cdots,n$$

After carrying out the integration, yields

$$[K(q,\mu)]\{q\} = \{R\} = 0 \tag{2.11}$$

where $[K(q,\mu)]$ is the total stiffness matrix, $\{R\} = [R_0^c,R_1^c,\cdots,R_n^c,R_1^s,\cdots,R_n^s]^T$ and

$$\{q\} = [\omega,a_1,a_3,\cdots,a_{2n+1},b_3,\cdots,b_{2n+1}]^T.$$

2.1.3.2 Analysis of oscillatory parametric regions

We assume that Eq. (2.7) exists undamped oscillation and the steady state response of form

$$u(\tau) = a\cos\tau, \omega = \omega_0 \tag{2.12}$$

where $|a|$ and ω_0 denote the amplitude and angular frequency of oscillation, respectively.

Substituting Eq. (2.12) into Eq. (2.7) and using the Galerkin procedure result in the following harmonic balance equations.

$$\omega_0^{1+\lambda}\sin\frac{\lambda\pi}{2} - 1 - \varepsilon\omega_0^\lambda\cos\frac{\lambda\pi}{2}(\frac{3}{4}a^2-1) = 0 \tag{2.13}$$

$$\omega_0^{1+\lambda}\cos\frac{\lambda\pi}{2} - \varepsilon\omega_0^\lambda\sin\frac{\lambda\pi}{2}(1-\frac{1}{4}a^2) = 0 \tag{2.14}$$

Simplifying Eqs. (2.13) and (2.14), we have

$$\omega_0 = \varepsilon(1-\frac{1}{4}a^2)\tan\frac{\lambda\pi}{2} \tag{2.15}$$

$$\cos\frac{\lambda\pi}{2} = \varepsilon\omega_0^\lambda(1-\frac{1}{4}a^2-\frac{1}{2}a^2\cos^2\frac{\lambda\pi}{2}) \tag{2.16}$$

Thus

$$\cos\frac{\lambda\pi}{2} = \varepsilon^{1+\lambda}\tan^{\lambda}\frac{\lambda\pi}{2}(1-\frac{1}{4}a^2)^{\lambda}(1-\frac{1}{4}a^2-\frac{1}{2}a^2\cos^2\frac{\lambda\pi}{2}) \tag{2.17}$$

To get positive real values of amplitude and angular frequency, the following conditions are needed

$$0<1-\frac{1}{4}a^2<1, \;\; 0<1-\frac{1}{4}a^2-a^2\cos^2\frac{\lambda\pi}{2}<1, \;\; \varepsilon^{1+\lambda}\tan^{\lambda}\frac{\lambda\pi}{2}>\cos\frac{\lambda\pi}{2}. \tag{2.18}$$

This result in a relation between control parameter ε and fractional order λ that indicates the condition in which the limit cycle can be generated by the system in (2.4):

$$\varepsilon \geq \frac{\sin(\lambda\pi/2)^{1/1+\lambda}}{\tan(\lambda\pi/2)} \tag{2.19}$$

which is identical to the result of Tavazoei [M.S. Tavazoei et al., 2009].

As a result, when Eq. (2.19) is satisfied, the system (2.4) might have undamped oscillation. Figure 2.1 shows the boundary between oscillatory and non-oscillatory regions in the (λ,ε) plane for the fractional van der Pol oscillator (2.4).

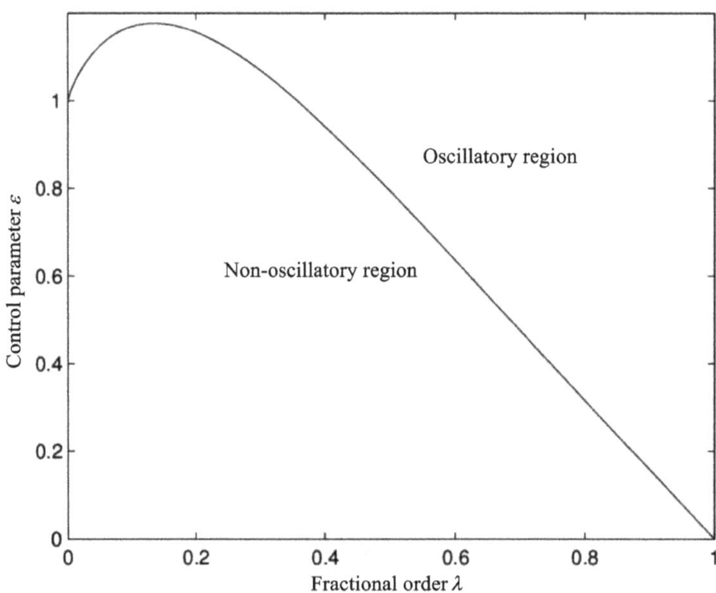

Figure 2. 1 Oscillatory and non-oscillatory regions of the system (2.4)

2.1.4 Residue Harmonic Balance Analysis

It should be pointed that it is rather difficult to calculate higher order harmonic balance equations. Thus, the above technique is only capable for obtaining the rough approximations to the angular frequency and limit cycle of Eq. (2.4). In this section, residue harmonic balance method is proposed to give an improvement to the above rough approximations. We assume the solutions to be of the form

$$u(\tau) = u_0(\tau) + pu_1(\tau) + p^2 u_2(\tau) + \cdots$$

$$\omega = \omega_0 + p\omega_1 + p^2 \omega_2 + \cdots \tag{2.20}$$

where p is an order parameter, $\omega_i (i = 0,1,2,\cdots)$ are unknowns to be determined and

$$u_0(\tau) = a_{1,0} \cos \tau,$$

$$u_k(\tau) = a_{1,k} \cos \tau + \sum_{i=1}^{k} \{a_{2i+1,k} \cos[(2i+1)\tau] + b_{2i+1,k} \sin[(2i+1)\tau]\}, k = 1,\cdots \tag{2.21}$$

2.1.4.1 The zeroth-order approximation

Substituting Eq. (2.20) into Eq. (2.7) and equating the coefficients of the p^0, yields

$$\Phi_0(\omega_0, u_0) \overset{\Delta}{=} \omega_0^{1+\lambda} D_\tau^{1+\lambda} u_0 + a\omega_0^{\lambda}(u_0^2 - 1)D_\tau^{\lambda} u_0 + u_0 = \sum_{i=1}^{2} [\Gamma_{2i-1,0} \cos(2i-1)\tau + \Upsilon_{2i-1,0} \sin(2i-1)\tau]$$

$$\tag{2.22}$$

Based on Galerkin procedure, the right hand side of Eq. (2.22) should not contain term $\cos(\tau)$, $\sin(\tau)$ because they can bring such so-called secular terms. Letting their coefficient be zeros, that is

$$\Gamma_{1,0}(\omega_0, a_{1,0}) = \frac{2}{\pi} \int_0^\pi \Phi_0 \cos \tau \, d\tau = 0$$

$$\Upsilon_{1,0}(\omega_0, a_{1,0}) = \frac{2}{\pi} \int_0^\pi \Phi_0 \sin \tau \, d\tau = 0 \tag{2.23}$$

The zeroth-order approximations of Eq. (2.4) can be obtained by solving the Eq. (2.23) with help of symbolic computations, as below

$$u_0(\tau) = a_{1,0}\cos\tau, \omega = \omega_0 \tag{2.24}$$

When substituting the obtained harmonic balance solution into Eq. (2.22), the terms $\Gamma_{3,0}$ and $\Upsilon_{3,0}$ are generally nonzero. We note these nonzero residuals as

$$\Phi_0 \overset{\Delta}{=} \Gamma_{1,0}\cos\tau + \Upsilon_{1,0}\sin\tau + \Gamma_{3,0}\cos 3\tau + \Upsilon_{3,0}\sin 3\tau = pR_0 \tag{2.25}$$

2.1.4.2 The first-order approximation

Substituting Eqs. (2.20), (2.25) into Eq. (2.7) and equating the coefficients of the p^1, gives

$$\Phi_1(\omega_1,u_1) \overset{\Delta}{=} \omega_0^{1+\lambda}D_\tau^{1+\lambda}u_1 + (1+\lambda)\omega_0^{\lambda}\omega_1 D_\tau^{1+\lambda}u_0 + \varepsilon\omega_0^{\lambda}(u_0^2-1)D_\tau^{\lambda}u_1 + \lambda\varepsilon\omega_0^{\lambda-1}\omega_1(u_0^2-1)D_\tau^{\lambda}u_0$$

$$+ 2\varepsilon\omega_0^{\lambda}u_0u_1 D_\tau^{\lambda}u_0 + u_1 + R_0 = \sum_{i=1}^{3}[\Gamma_{2i-1,1}\cos(2i-1)\tau + \Upsilon_{2i-1,1}\sin(2i-1)\tau] \tag{2.26}$$

Based on Galerkin procedure, the right hand side of Eq. (2.26) should not contain term $\cos(\tau)$, $\sin(\tau)$, $\cos(3\tau)$ and $\sin(3\tau)$ because they can bring such so-called secular terms. Letting their coefficient be zeros, that is

$$\Gamma_{2i-1,1}(\omega_1,a_{1,1},a_{3,1},b_{3,1}) = \frac{2}{\pi}\int_0^{\pi}\Phi_1\cos[(2i-1)\tau]d\tau = 0, i = 1,2,$$

$$\Upsilon_{2i-1,1}(\omega_1,a_{1,1},a_{3,1},b_{3,1}) = \frac{2}{\pi}\int_0^{\pi}\Phi_1\sin[(2i-1)\tau]d\tau = 0, i = 1,2. \tag{2.27}$$

The first-order approximations of Eq. (2.4) can be obtained by solving the linear Eqs. (2.27) as below

$$u_{(1)}(\tau) = (a_{1,0}+a_{1,1})\cos\tau + a_{3,1}\cos 3\tau + b_{3,1}\sin 3\tau, \omega_{(1)} = \omega_0 + \omega_1 \tag{2.28}$$

When substituting the obtained the first-order approximations (2.28) into Eq. (2.22), the terms $\Gamma_{5,1}$ and $\Upsilon_{5,1}$ are generally nonzero. We note nonzero residuals as

$$\Phi_1 \overset{\Delta}{=} \sum_{i=1}^{3}\{\Gamma_{2i-1,1}\cos[(2i-1)\tau] + \Upsilon_{2i-1,1}\sin[(2i-1)\tau]\} = p^2R_1 \tag{2.29}$$

2.1.4.3 The second-order approximation

Substituting Eqs. (2.20) and (2.29) into Eq. (2.7) and equating the coefficients of the p^2, gives

$$\Phi_2(\omega_2, u_2) \overset{\Delta}{=} \omega_0^{1+\lambda} D_\tau^{1+\lambda} u_2 + (1+\lambda)\omega_0^\lambda \omega_1 D_\tau^{1+\lambda} u_1 + [(1+\lambda)\omega_0^\lambda \omega_2 + \frac{\lambda(\lambda+1)}{2}\omega_0^{\lambda-1}\omega_1^2] D_\tau^{1+\lambda} u_0$$

$$+ \varepsilon\omega_0^\lambda (u_0^2 - 1) D_\tau^\lambda u_2 + \varepsilon\omega_0^{\lambda-1} D_\tau^\lambda u_1 [2\omega_0 u_0 u_1 + \omega_1(u_0^2 - 1)] + 2\varepsilon\lambda\omega_0^{\lambda-1}\omega_1 u_0 u_1 D_\tau^\lambda u_0$$

$$+ \varepsilon\omega_0^\lambda (2u_0 u_2 + u_1^2) D_\tau^\lambda u_0 + \varepsilon[\lambda\omega_0^{\lambda-1}\omega_2 + \frac{\lambda(\lambda-1)}{2}\omega_0^{\lambda-2}\omega_1^2](u_0^2 - 1) D_\tau^\lambda u_0 + u_2 + R_1$$

$$= \sum_{i=1}^{4} [\Gamma_{2i-1,2} \cos(2i-1)\tau + \Upsilon_{2i-1,2} \sin(2i-1)\tau]$$

$$(2.30)$$

The right hand side of Eq. (2.30) should not contain term $\cos(\tau)$, $\sin(\tau)$, $\cos(3\tau)$, $\sin(3\tau)$, $\cos(5\tau)$ and $\sin(5\tau)$ because they can bring such so-called secular terms. Letting their coefficient be zeros, that is

$$\Gamma_{2i-1,2}(\omega_2, a_{1,2}, a_{3,2}, b_{3,2}, a_{5,2}, b_{5,2}) = \frac{2}{\pi}\int_0^\pi \Phi_2 \cos[(2i-1)\tau]\mathrm{d}\tau = 0, i = 1, 2, 3,$$

$$(2.31)$$

$$\Upsilon_{2i-1,2}(\omega_2, a_{1,2}, a_{3,2}, b_{3,2}, a_{5,2}, b_{5,2}) = \frac{2}{\pi}\int_0^\pi \Phi_2 \sin[(2i-1)\tau]\mathrm{d}\tau = 0, i = 1, 2, 3.$$

The second-order approximations of Eq. (2.4) can be obtained by solving the Eqs. (2.31) as below

$$u_{(2)}(\tau) = \sum_{i=0}^{2} a_{1,i} \cos\tau + \sum_{i=1}^{2} (a_{3,i} \cos 3\tau + b_{3,i} \sin 3\tau) + a_{5,2} \cos 5\tau + b_{5,2} \sin 5\tau, \omega_{(2)} = \sum_{i=0}^{2} \omega_i$$

$$(2.32)$$

It should be clear how the procedure works for constructing further higher-order approximations to the angular frequency and limit cycle. It should be noted that Eqs. (2.27), (2.31) etc are linear with respect to unknowns ω_j, $a_{1,j}, \cdots, a_{2j+1,j}$, $b_{3,j}, \cdots, b_{2j+1,j}$, $j=1,2,\cdots$, which implies that we can obtain the solutions to any desired accuracy rather easily.

2.1.5 Analytically Asymptotic Solution of Steady State Response

2.1.5.1 Asymptotic solution under different fractional orders

In this subsection, we illustrate the effects of the fractional order on the angular frequency and amplitude of limit cycle and give out the comparison between asymptotic approximations and numerical results. According to Eq. (2.19) and Figure2.1, it can be obtained that the system (2.4) might have undamped oscillation when fractional order $\lambda > 0.35692$ for control parameter $\varepsilon = 1$. Note that the zeroth-order harmonic approximation to the angular frequency ω_0 and amplitude $a_{1,0}$ of limit cycle can be obtained by the software Maple or Mathematics. Due to the higher order approximations of residue harmonic balance method are governed by linear equations, thus solutions can be obtained to any desired accuracy without difficulty. The fractional order versus frequency and fractional order versus amplitude curves of limit cycles are shown in Figs. 2.2 and 2.3 respectively. From Figs. 2.2 and 2.3, we can see that the angular frequency of limit cycle in steady state increases at first, and then decreases for the fractional order varies from 0.357 to 1. However, the amplitude of limit cycle in steady state always increases. It is also clear that the second-order approximations are high accuracy and in good agreement with numerical results.

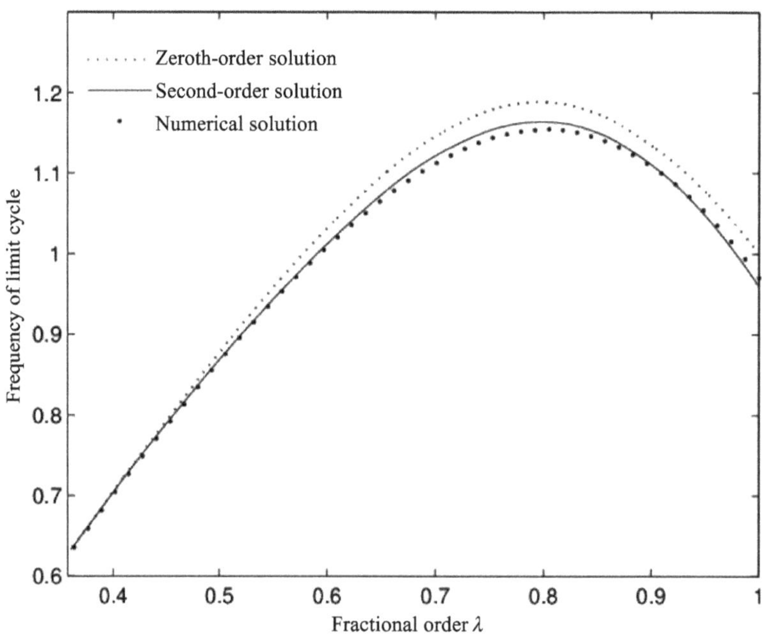

Figure.2. 2 Comparisons of frequency and effect of fractional order

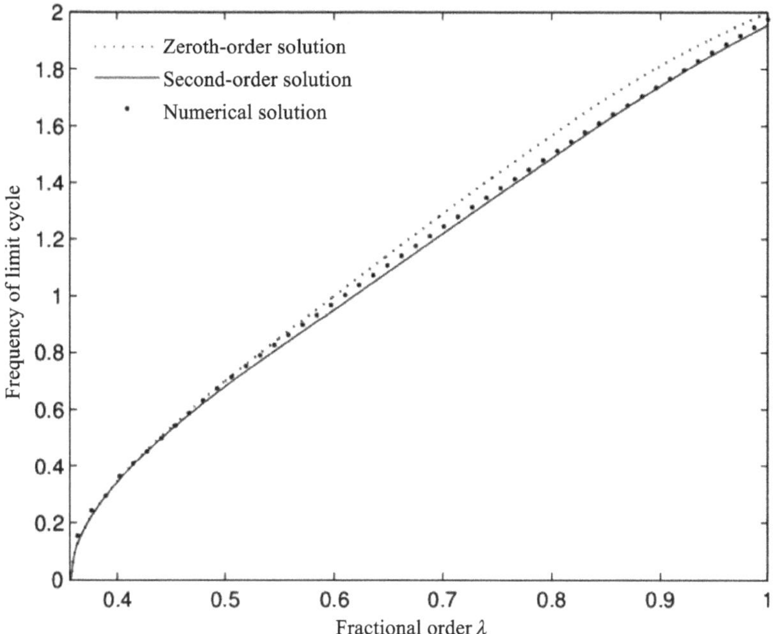

Figure.2. 3 Comparison of amplitude and effect of fractional order

2.1.5.2 Asymptotic solution under different control parameter

In this subsection, we take $\lambda = 0.5$ in our fractional derivative. Then Eq. (2.4) becomes

$$D_t^{1.5}u(t) + \varepsilon[u(t)^2 - 1]D_t^{0.5}u(t) + u(t) = 0 \tag{2.33}$$

According to Eq. (2.19), it is evident that the system (2.33) might have undamped oscillation when control parameter $\varepsilon > 0.7937$. Figs. 2.4 and 2.5 give the comparison between the asymptotic approximations to the angular frequency and amplitude of limit cycle and corresponding numerical results. It is clear that the first-order approximations are high accuracy. As the control parameter ε increases from 0.7937 to 2, the frequency and amplitude of limit cycle increase. And the amplitude is close to zero when $\varepsilon \approx 0.7937$. To further illustrate and verify the accuracy and effectiveness of the residue harmonic balance approximations, the comparison of the phase portraits derived are presented in Figure 2.6 for $\varepsilon \approx 1, 1.5, 2$ respectively, in which solid lines and dots denote the first-order analytical solution and numerical result, respectively. As we can see that, the first-order approximations are in good agreement with the numerical results.

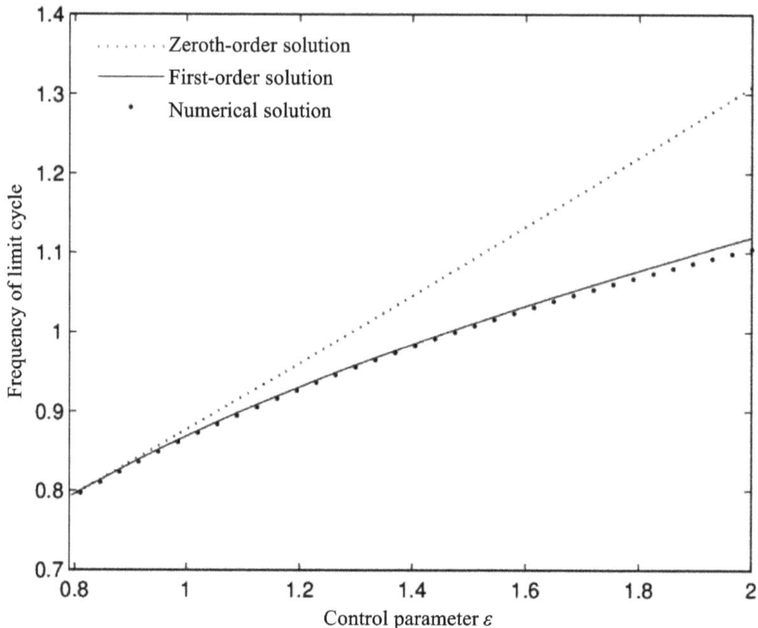

Figure 2. 4 Comparisons of frequency and effect of control parameter

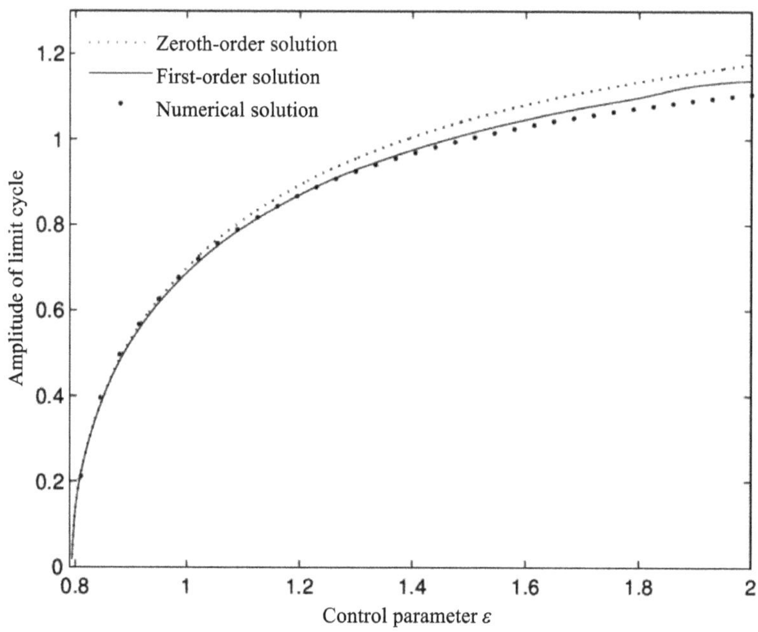

Figure 2. 5 Comparisons of amplitude and effect of control parameter

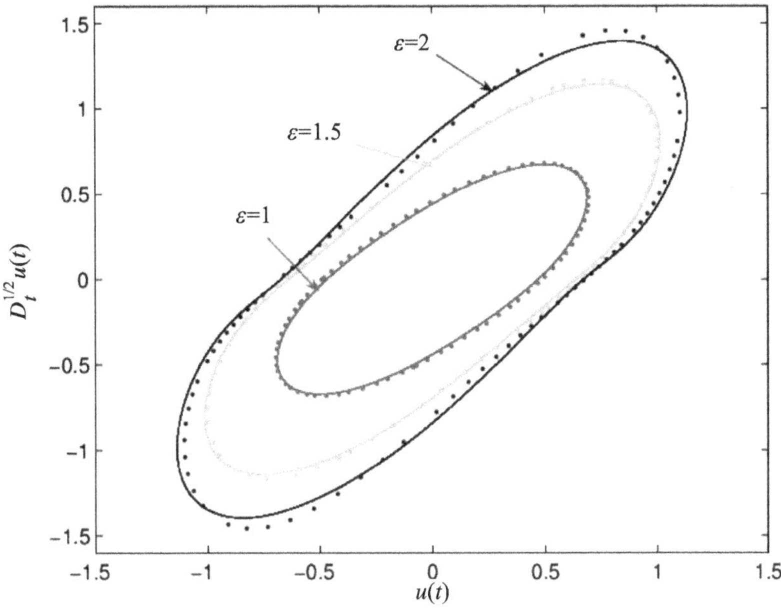

Figure 2. 6 Phase portraits of limit cycle with the control parameters ε ={1,1.5,2}

2.1.6 Concluding Remarks

In this section, the residue harmonic balance is extended to study the fractional van der Pol oscillator. The oscillatory parametric region of the oscillator has been predicted and given out. The result is consistent with those of the previously introduced method based on stability theorems. The highly accurate asymptotic solutions of steady state are obtained and compared. Moreover, the effects of the fractional order on the angular frequency and amplitude of limit cycle and the asymptotic solutions of different control parameter with one-half derivative are also analyzed. The results show that residue harmonic balance method has great potential to fractionally damped systems.

2.2 Asymptotic Solution of an Autonomous Duffing–van der Pol Oscillator with Fractional Derivative

2.2.1 Introduction

Lifshitz et al [R. Lifshitz et al., 2008, 2011] considered a single resonator, which is driven parametrically by modulating its effective stiffness and the corresponding autonomous equation is given by

$$\ddot{u} + (h_1 + h_2 u^2)\dot{u} + \Omega_0^2 u + ku^3 = 0 \tag{2.34}$$

Only a minor change is given in our equation, namely the extension of damping term by introducing fractional derivative. The equation of motion is then given by

$$\ddot{u} + (h_1 + h_2 u^2)D_t^{\alpha} u + \Omega_0^2 u + ku^3 = 0 \tag{2.35}$$

where D_t^{α} is the fractional derivative in the sense of Caputo definition.

Recently, some similar researches can be found. For instance, Xie and Lin [F. Xie & X.Y. Lin, 2009] introduced the following van der Pol oscillator with a small fractional damping term and studied the asymptotic solution of Cauchy problem.

$$\ddot{u} - \varepsilon[1 - u^2(t)]D_t^{\lambda}u + u(t) = 0$$

In ref [A.Y.T. Leung et al., 2012], Leung et al studied a modified version of fractional van der Pol oscillator by considering a fractional restoring force, reads

$$\ddot{u} - \varepsilon[1 - u^2(t)]D_t^{\lambda}u + u^{1/3}(t) = 0$$

In this section, we analyzes the influences of fractional order and cubic stiffness on steady state responses for Eq. (2.35) by obtaining higher-order approximate solutions to angular frequency and amplitude via residue harmonic balance method.

2.2.2 Residue Harmonic Balance Equation of Various Order Approximations

Since the period is unknown in Eq. (2.35). We can introduce a normalized period of 2π by introducing a transformation $x = \omega t$ where ω is an unknown frequency, then Eqs. (2.35) can be rewritten

$$\omega^2 u'' + (h_1 + h_2 u^2)\omega^\alpha D_x^\alpha u + \Omega_0^2 u + ku^3 = 0 \tag{2.36}$$

For Eq. (2.36), according to the residue harmonic balance procedure, the zeroth-, first- and second-order solutions can be obtained. The corresponding residue harmonic balance equation is placed in Appendix B.

2.2.3 Results and Discussions

2.2.3.1 Asymptotic solution of resonator under different fractional orders

This subsection illustrates the variation of the response to angular frequency and amplitude under different fractional order derivatives.
The parameters $h_1=-1$, $h_2=1$, $\Omega_0=1$, $k=1$ are applied and fractional order α is a bifurcation parameter. The relation of fractional order versus stable response to frequency and amplitude are shown in Figures 2.7 and 2.8 respectively. The zeroth-, first- and second-order solutions are indicated, respectively, by the dashed, dash-dot and solid lines. The dots correspond to results of numerical simulation. From Figures 2.7 and 2.8, it is easily seen that the second-order approximate frequency of steady response increases slightly at first, and then decreases for the fractional order from 0 to 1. However, the second-order approximate amplitude decreases at first until fractional order $\alpha \approx 0.84$, and then increases for the fractional order from 0.84 to 1. It is also clear that the second-order approximation is high accuracy and in good agreement with numerical result. However, the zeroth-order approximation gives an error prediction. To show the efficiency of the residue harmonic balance method, a comparison about the frequency and amplitude of the

various-order approximations is tabulated in Table 2.1.

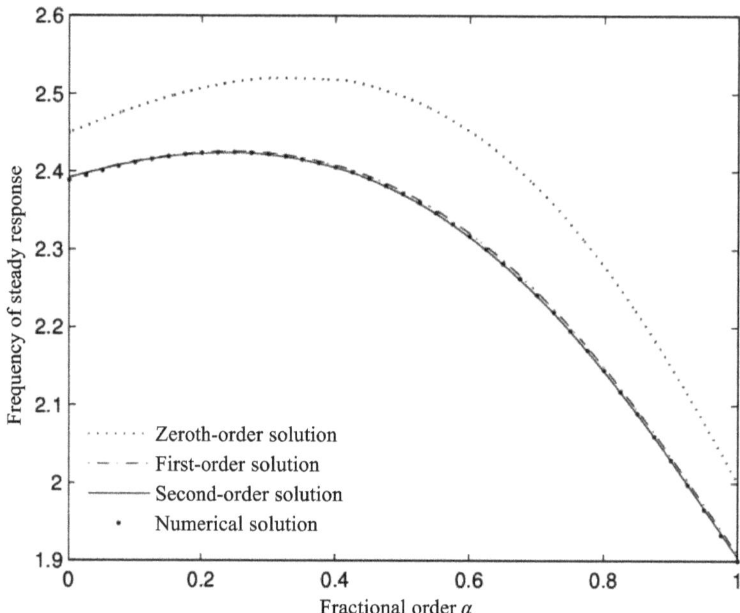

Figure 2.7 The present angular frequencies as function of fractional order α and comparison with numerical result for Eq. (2.35)

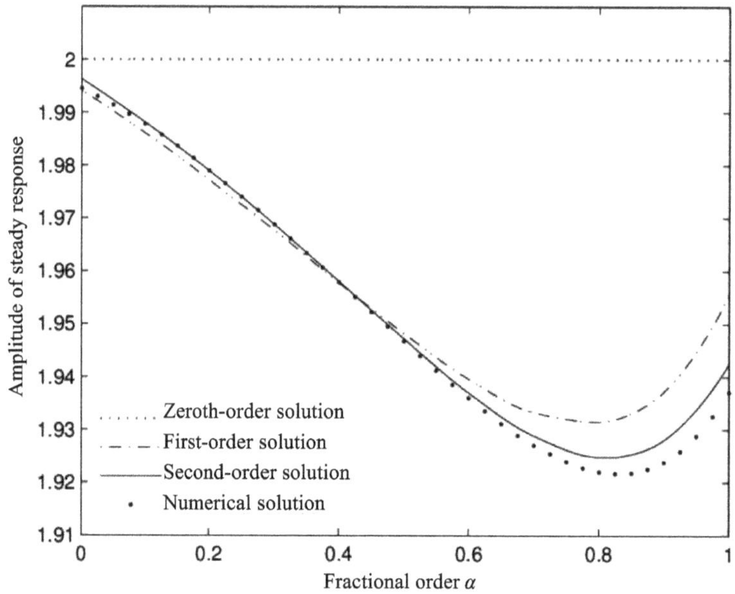

Figure 2. 8 The present amplitudes as function of fractional order α and comparison with numerical result for Eq. (2.35)

Table 2. 1 Comparison of the approximate results and numerical solutions for Eq. (2.35)
under different fractional orders

α	$\omega_{(0)}$	$\omega_{(1)}$	$\omega_{(2)}$	$\omega_{(N)}$	$A_{(0)}$	$A_{(1)}$	$A_{(2)}$	$A_{(N)}$
0.1	2.4826	2.4129	2.4123	2.4117	2	1.9861	1.9881	1.9878
0.2	2.5072	2.4244	2.4232	2.4237	2	1.9772	1.9790	1.9790
0.4	2.5182	2.4081	2.4050	2.4057	2	1.9578	1.9579	1.9579
0.6	2.4523	2.3213	2.3162	2.3173	2	1.9396	1.9370	1.9360
0.8	2.2792	2.1496	2.1441	2.1450	2	1.9315	1.9249	1.9220
1.0	2.0000	1.9095	1.9048	1.9002	2	1.9556	1.9425	1.9371

2.2.3.2 Asymptotic solution of resonator under different cubic stiffnesses

In this subsection, we take $\alpha = 0.5$ in our fractional derivative. Figure2.9 gives a comparison betwcen the asymptotic approximations to the angular frequency and amplitude of steady responses and corresponding numerical results. The parameters $h_1 = -1$, $h_2 = 1$, $\Omega_0 = 1$ are applied and cubic stiffness k is the bifurcation parameter. It is evident that the residue harmonic balance analysis yields better solutions on the steady response to angular frequency and amplitude. Table 2.2 gives a comparison about the present zeroth-, first- and second-order approximate frequency, amplitude and numerical results. We can see that the present second-order solution is in excellent agreement with numerical solution. Moreover, the present second-order approximate frequency keeps increasing all time for cubic stiffness from 0 to 10, the second-order approximate amplitude changes a little.

**Table 2. 2 Comparison of the approximate results and numerical solutions for Eq. (2.35)
with $\alpha = 0.5$ under different cubic stiffnesses**

k	$\omega_{(0)}$	$\omega_{(1)}$	$\omega_{(2)}$	$\omega_{(N)}$	$A_{(0)}$	$A_{(1)}$	$A_{(2)}$	$A_{(N)}$
0.1	1.7861	1.6913	1.6918	1.6925	2	1.9412	1.9398	1.9389
1.0	2.4969	2.3748	2.3706	2.3714	2	1.9482	1.9472	1.9467
2.0	3.0792	2.9356	2.9295	2.9306	2	1.9517	1.9509	1.9505
4.0	3.9775	3.80115	3.7928	3.7940	2	1.9552	1.9544	1.9542
6.0	4.6973	4.4948	4.4849	4.4859	2	1.9569	1.9562	1.9560
8.0	5.3161	5.0910	5.0797	5.0814	2	1.9580	1.9573	1.9572
10	5.8673	5.6221	5.6097	5.6097	2	1.9558	1.9580	1.9579

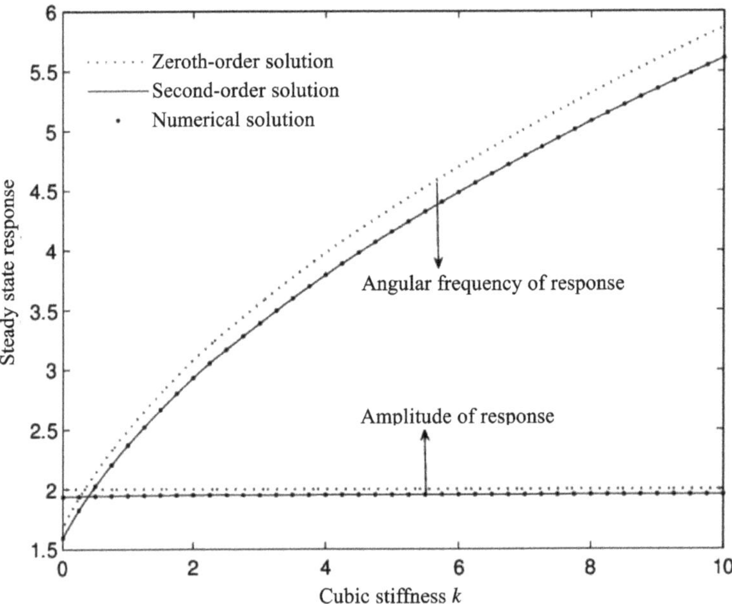

**Figure 2. 9 The present frequencies and amplitudes as function of cubic stiffness and
comparison with numerical result for Eq. (2.35)**

2.2.4 Concluding Remarks

In this section, the damping term of an autonomous Duffing-van der Pol oscillator
is extended by introducing the fractional derivative. Then the asymptotic solutions

of the fractional oscillator are obtained and compared by using residue harmonic balance approach. The results show that the present second-order approximations are in good agreement with respect to numerical results. In addition, the influences of fractional order and cubic stiffness on steady state responses are illustrated and discussed.

Chapter 3 Resonance Response and Bifurcation of Periodically Excited Systems Containing Fractional Derivative

3.1 Resonance Response and Analytical Approximation of a Duffing Oscillator with Fractional–order Derivative

3.1.1 Introduction

In the past decades, fractional differential equations are increasingly applied in modeling many physical and engineering problems [I. Podlubny, 1999; L. Debnath, 2003; Y.A. Rossikhin & M.V. Shitikova, 2004, 2010]. It has been found that the behavior of many dynamic systems in various fields including thermal engineering, acoustics, visoelasticity, electromagnetism, control, turbulence, robotics, signal processing and some other physical processes can be properly described by fractional order system theory. On the one hand, the response and dynamical behavior of the fractional differential systems has started to attract attention in the last few decades. One of the reasons of studying fractional oscillators is the study of vibrations of more complex structures can be reduced to vibrations of a set oscillators [Y.A. Rossikhin & M.V. Shitikova, 2004, 2010]. Many phenomena are found in integer order dynamic system, such as resonance, hysteresis, symmetry breaking, bifurcation and chaos can be also exhibited in fractional differential system [Y.A. Rossikhin & M.V. Shitikova, 2004, 2010; R.S. Barbosa et al., 2007; X. Gao & J. Yu, 2005; Y.G. Yu et al., 2009]. On the other hand, owing to exact solutions of most of the nonlinear fractional differential systems cannot be found easily. Thus finding accurate and efficient methods for solving fractional differential systems has been an active research undertaking. Some numerical and analytically approximate methods were presented [S.S. Ray et al., 2005; Z.M. Odibat, 2010; M. Surigat et al., 2010; A.A. Rabtah et al., 2010]. In this section, we introduce the method of harmonic balance in combination with polynomial continuation to find the steady state responses of Duffing oscillator with fractional order derivative and examine the influence of fractional representations on the response behaviors. In addition, some nonlinear phenomena of fractional order oscillator, such as resonance, supercritical pitchfork bifurcation, symmetry breaking, hysteresis and jump are also investigated analytically.

3.1.2 Duffing Oscillator with Fractional Order Derivative

In this section, we consider a Duffing type oscillator with damping defined by fractional derivative, which is given by

$$m\ddot{u}(t) + hD_t^\alpha u(t) + ku(t) + ru(t)^3 = f\cos\omega t \tag{3.1}$$

where u, m, h, k, f and ω are respectively the displacement, mass, damping coefficient, linear stiffness, cubic stiffness of the oscillator and the excitation amplitude and frequency. D_t^α is the fractional derivative of $u(t)$ of order $\alpha(0 < \alpha < 2)$ and in the sense of Caputo definition is defined as

$$D_t^\alpha u(t) = \frac{1}{\Gamma(m-\alpha)} \int_0^t (t-\tau)^{m-\alpha-1} \frac{\mathrm{d}^m}{\mathrm{d}\tau^m} u(\tau)\,\mathrm{d}\tau \tag{3.2}$$

for $m-1 < \alpha \le m, m \in N, t > 0, u \in C_{-1}^m$.

Eq. (3.1) has been widely studied. For instance, Sheu et al [L.J. Sheu et al., 2007] examined the chaotic dynamics of the fractionally damped Duffing equation using predictor-corrector method. He [J.H. He, 1998] applying variational iteration method to obtain approximate analytical solution for case of $f = 0$ and half-order Ricmann-Liouville fractional derivative. Padovan and Sawicki [J. Padovan & J.T. Sawicki, 1998] utilized the energy constrained Lindstedt-Poincare perturbation procedure to establish the harmonic solution of fractionally damped systems by employing a Diophantine version of the fractional operator powers.

3.1.3 Steady State Response and Harmonic Balance Method

The steady state response of Eq. (3.1) can be expanded in Fourier series

$$u_0(t) = \frac{a_0}{2} + \sum_{i=1}^{n} a_i \cos(i\omega t + \phi_i) \tag{3.3}$$

where a_0 is a bias term, $a_i, (i = 1,2,\cdots)$ and $\phi_i, (i = 1,2,\cdots)$ are the amplitudes and phase angle respectively. n is an integer representing the maximum order which retained harmonics. For convenience, we introduce a new time scale $\tau = \omega t$, then Eqs. (3.1) and (3.3) become respectively

$$m\omega^2 u''(\tau) + h\omega^\alpha D_\tau^\alpha u(\tau) + ku(\tau) + ru(\tau)^3 = f\cos\tau \tag{3.4}$$

$$u_0(\tau) = \frac{a_0}{2} + \sum_{i=1}^{n} a_i \cos(i\tau + \phi_i) \tag{3.5}$$

Substituting Eq.(3.5) into Eq.(3.4) and using the Galerkin procedure and expression Eq. (3.2) of fractional order derivative, result in the following harmonic balance equations.

$$R_i^c(a_0,a_1,\cdots,a_n,\phi_1,\cdots,\phi_n) = \frac{2}{\pi}\int_0^\pi (m\omega^2 u'' + h\omega^\alpha D_\tau^\alpha u + ku + ru^3 - f\cos\tau)\cos i\tau\,d\tau = 0, i = 0,\cdots,n$$

$$R_i^s(a_0,a_1,\cdots,a_n,\phi_1,\cdots,\phi_n) = \frac{2}{\pi}\int_0^\pi (m\omega^2 u'' + h\omega^\alpha D_\tau^\alpha u + ku + ru^3 - f\cos\tau)\sin i\tau\,d\tau = 0, i = 1,\cdots,n$$

Carrying out the integration, yields

$$[K_1(q_1,\mu)]\{q_1\} - \{F\} = \{R_1\} = 0 \tag{3.6}$$

where $[K_1(q_1,\mu)]$ is the total stiffness matrix, $\{R_1\} = [R_0^c, R_1^c, \cdots, R_n^c, R_1^s, \cdots, R_n^s]^T$, $\{q_1\} = [2a_0, a_1\cos\phi_1, \cdots, a_n\cos\phi_n, -a_1\sin\phi_1, \cdots, -a_n\sin\phi_n]^T$ and $\{F\}$ is the Fourier coefficient vector of the excitation. For given the values of μ, Eq. (3.6) is solved using the method of harmonic balance in combination with homotopy continuation algorithm. Further, all the steady state analytical solutions are obtained.

3.1.4 The Analysis of Resonance Response and Discussion

Due to the higher order harmonic terms make little contribution to the steady state solutions of Eq. (3.1). We will consider the third-order symmetric harmonic and second-order asymmetric harmonic cases, and the steady state solutions can be obtained by harmonic balance in combination with polynomial homotopy continuation. The influence of the excitation frequency and fractional damping of fractional Duffing oscillator are examined by the fractional order versus amplitudes, fractional order versus phase angle and frequency responses amplitude curves. Some nonlinear phenomena, such as superharmonic resonance, supercritical pitchfork bifurcation, hysteresis and jump phenomena are investigated and shown. In addition, the steady state analytical solutions are given and compared to verify the validity of the present method to fractional differential equation.

3.1.4.1 Effect of excitation frequency

Figure 3.1 illustrates the effects of the excitation frequency on the fractional order versus amplitudes and fractional order versus phase angle responses. The steady state symmetric approximate solution is assumed to be

$$u(t) = a_1 \cos(\omega t + \phi_1) + a_3 \cos(3\omega t + \phi_3) \tag{3.7}$$

The corresponding fractional order versus amplitudes and phase angle curves of the steady state response are shown with parameters $m = 1$, $h = 0.4$, $k = 1$, $r = 1$, $f = 0.4$ in fractional Duffing type oscillator (3.1) in Figure 3. 1. From Figure3. 1a can be seen that an increasing of the excitation frequency ($\omega < 1.17$) yields an increasing in amplitudes of steady state response, where $A_i = |a_i|, i = 1,3$. And, the amplitudes increase with the increasing of fractional order. The fractional orders versus phase angle ϕ_1 curves for excitation frequency $\omega = 0.2, 0.5, 0.8, 1$ are shown in Figure 3.1b. The phase angles decrease at first, and then increase for the fractional order varies from 0 to 2. In addition, the phase angle is bigger for smaller excitation frequency.

For higher values of the excitation frequency, the fractional order versus amplitudes curves bend several times (Figure 3.1c), so that jump-up and jump-down points occur. From Figure 3.1a, a single jump occurs at α_3 point ($\alpha \approx 1.4648$) for excitation frequency $\omega = 1.12$. Figure 3.1c shows that the fractional order versus response amplitudes A_1 and $40A_3$ curves for $\omega = 1.5$, we can see that when fractional order increases to α_1 point ($\alpha \approx 0.3962$), the steady state responses bifurcate a pair of new solutions and low branch is stable, upper branch is unstable. At α_2 point ($\alpha \approx 0.538036$), a jump-down point occurs. With the fractional order continuously increases to α_4 point ($\alpha \approx 1.813194$), a jump-up point occurs, meanwhile two stable solutions coexist. From Figure 3.1c, we can find that the response amplitudes of A_1 and A_3 have similar shapes but the size represented by the vertical scales is quite different.

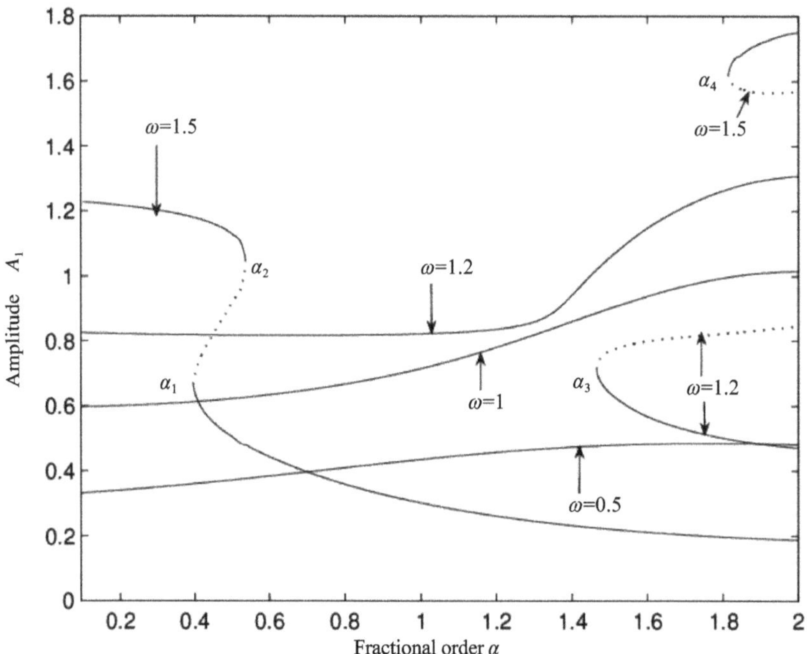

Figure 3.1a Fractional order versus response amplitude A_1 curves for case $\omega = 0.5, 1, 1.2, 1.5$

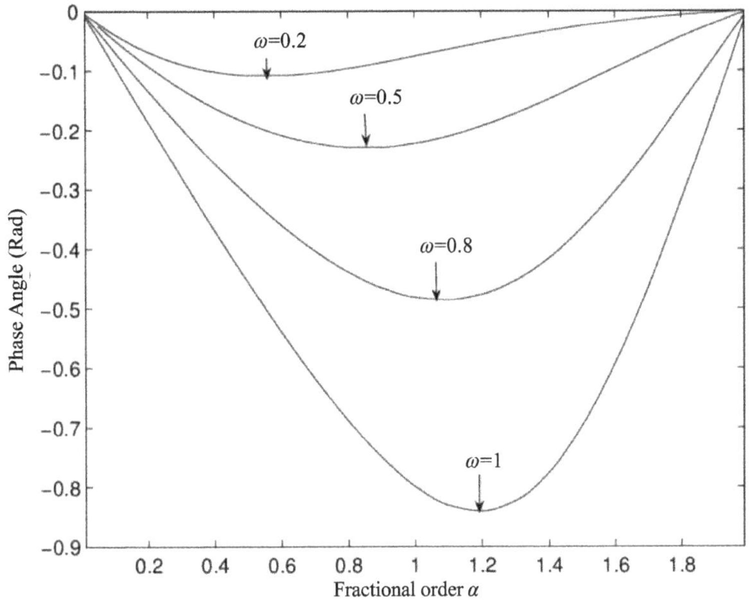

Figure 3.1b Fractional order versus phase angle ϕ_1 curves for case ω=0.2,0.5,0.8,1

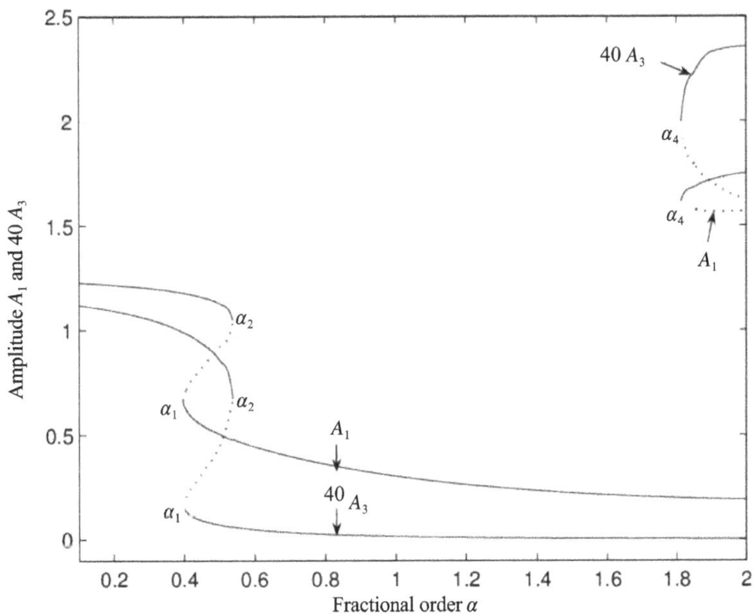

Figure 3.1c Fractional order versus response amplitudes $A1$ and $40A3$ curves for ω=1.5

3.1.4.2 Effect of fractional order on the frequency amplitude response

Figure 3.2 is presented to illustrate the effects of the fractional order on the frequency response curves. The steady state approximate solution is assumed to be Eq. (3.7). The corresponding frequency amplitude curves are shown with parameters $m = 1, h = 0.4, k = 1, r = 1, f = 0.4$ in Figure 3.2, the solid lines and broken lines correspond to stable and unstable states, respectively. From Figure 3.2, we find that the amplitudes A_1 and A_3 at first increase with the increasing of the excitation frequency from 1, then the amplitudes appear two jump phenomena which take place at points $\omega_1(\omega \approx 1.19775)$ and $\omega_2(\omega \approx 1.22216)$, $\omega_3(\omega \approx 1.473068)$ and $\omega_5(\omega \approx 1.54007)$, $\omega_4(\omega \approx 1.52748)$ and $\omega_6(\omega \approx 2.6348)$ corresponding to the fractional order $\alpha = 1.5, 0.5, 0.2$ respectively. Slowly increases the excitation frequency and beyond the points $\omega_2, \omega_5, \omega_6$ to cause jump and the amplitudes of response stay in lower branch state with the continually increasing of the excitation frequency. Although there are the similar shapes of amplitudes under different damping with fractional order $\alpha = 1.5, 0.5, 0.2$, the frequency amplitude relation possesses bigger differentiation that has been shown in Figure 3.2a and 3.2b. It should be pointed that the response amplitudes of A_1 and A_3 in steady state have similar shapes in Figure 3.2a and 3.2b but the size represented by the vertical scales are quite different.

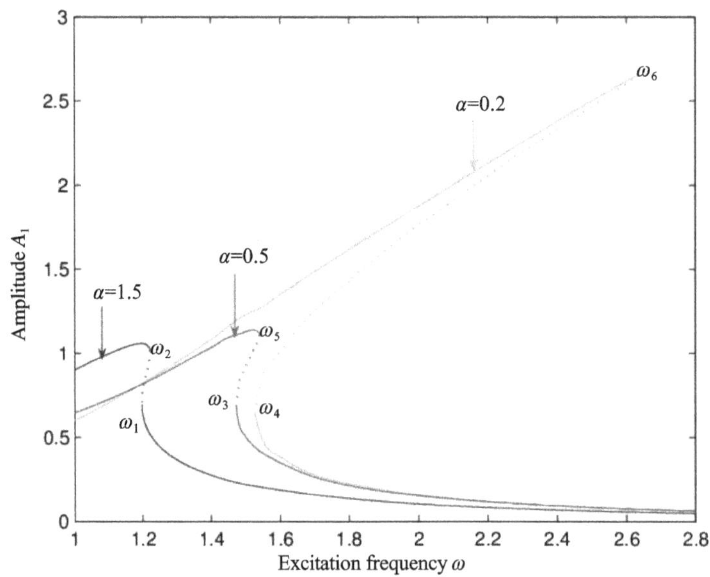

Figure 3.2a Steady state frequency amplitude A_1 curves
for fractional order $\alpha = 0.2, 0.5, 1.5$

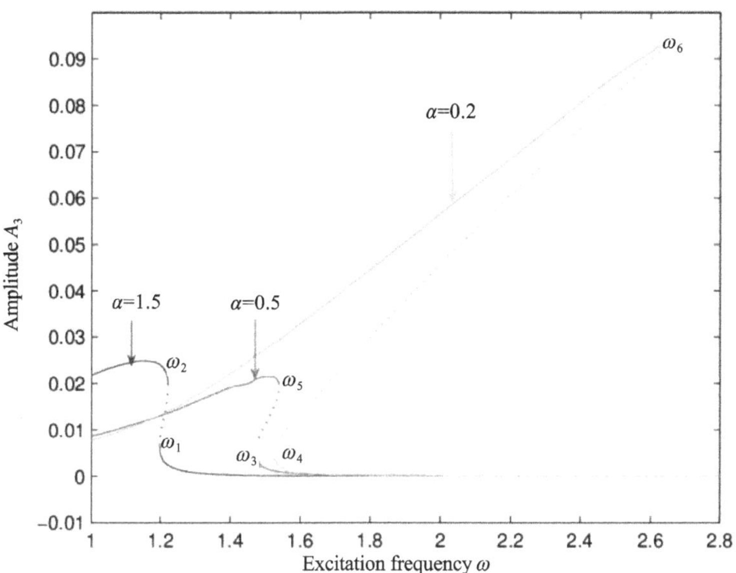

Figure 3.2b Steady state frequency amplitude A_3 curves

for fractional order α =0.2,0.5,1.5

3.1.4.3 Symmetry breaking bifurcation

In this subsection, an approximate solution corresponding to the steady state response in the region of the second order superharmonic resonance is sought. The approximate solution is assumed to be

$$u(t) = \frac{a_0}{2} + a_1 \cos(\omega t + \phi_1) + a_2 \cos(2\omega t + \phi_2)$$

The corresponding frequency amplitude curves of the steady state response with parameters $m = 1$, $h = 0.04$, $k = -0.2$, $r = 8/15$, $f = 0.4$ in one-half fractional Duffing type oscillator (3.1). In Figure 3.3, the solid lines and broken lines correspond to stable and unstable states, respectively. From Figure 3.3, we find two branch points $\omega_1(\omega \approx 0.7615975)$ and $\omega_3(\omega \approx 4.297)$, hysteresis effects, that is, the amplitudes appear two jump phenomena which take place at points ω_1 and ω_3. Starting with an oscillation represented by a point on the upper branch for less than point ω_3, the amplitude is "large". Slowly increase the excitation frequency (beyond the point ω_3) to cause a jump to the lower branch, that is, to small amplitude. A jump from small amplitude to large amplitude occurs at ω_1 point. And the upper branches are symmetric solutions. When excitation frequency increases to point $\omega_2(\omega \approx 0.8578)$, a symmetry breaking occurs, the original symmetric solution loses stable and a pair of stable asymmetric solution appear. (i) Just to the right of point ω_2, there are three solutions, an unstable symmetric solution and two stable asymmetric solutions. (ii) These solutions merge at point ω_2. (iii) Just to the left of point ω_2, there is single stable solution.

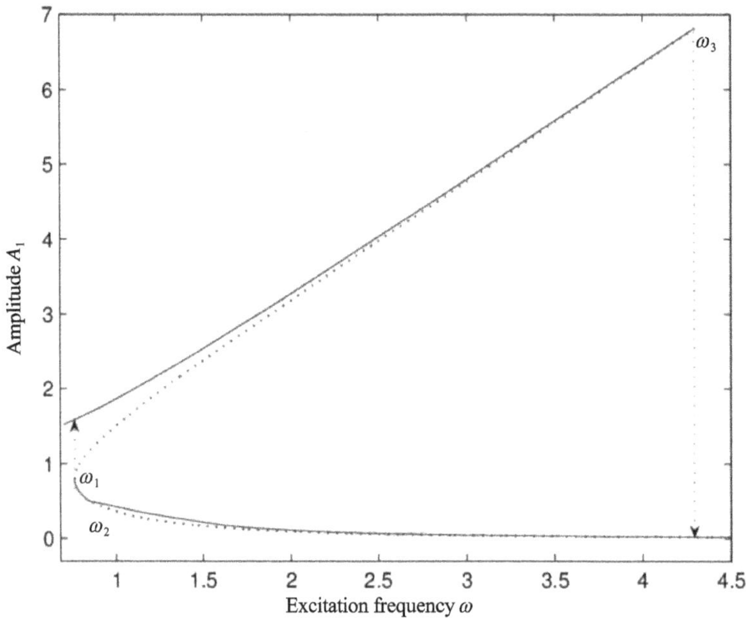

Figure3.3 Steady state frequency response curve for $\alpha =0.5$

3.1.4.4 Analytical solutions and comparison of phase portraits

For verifing the validity of harmonic balance method to fractional order system, we show that comparisons between analytical solutions and numerical results for parameter $m = 1, h = 0.4, k = 1, r = 1, f = 0.4, \omega = 1$ and fractional order $\alpha=0.2,0.5,0.8,1,1.2,1.5,1.8$. The analytical expressions for steady state response are tabulated in Table 3.1. Figure 3.4a and 3.4b show the phase portraits of corresponding solutions, the solid lines denote the analytical solution and starts are from numerical results. It is shown that analytical limit cycles are in excellent agreement with numerical results.

Table.3. 1 The analytical expressions for steady state response of Eq. (3.1)

	$\alpha=0.2$	$\alpha=0.5$	$\alpha=0.8$	$\alpha=1$	$\alpha=1.2$	$\alpha=1.5$	$\alpha=1.8$
$\cos(\omega t)$	0.59009	0.55856	0.51505	0.49845	0.52023	0.68998	0.9449
$\sin(\omega t)$	0.11151	0.27395	0.42411	0.51246	0.58124	0.57852	0.31034
$\cos(3\omega t)$	0.006654	0.00234	-0.00375	-0.00766	-0.0107	-0.00739	0.017322
$\sin(3\omega t)$	0.003973	0.008344	0.009772	0.009821	0.011141	0.020479	0.019955

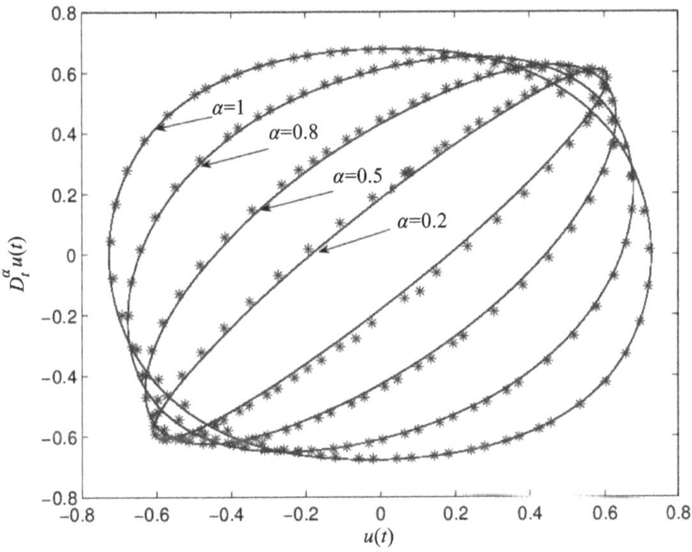

Figure3.4a Phase portraits for fractional order $\alpha = \{0.2,0.5,0.8,1\}$.

Solid line: analytical solution, Stars: numerical result

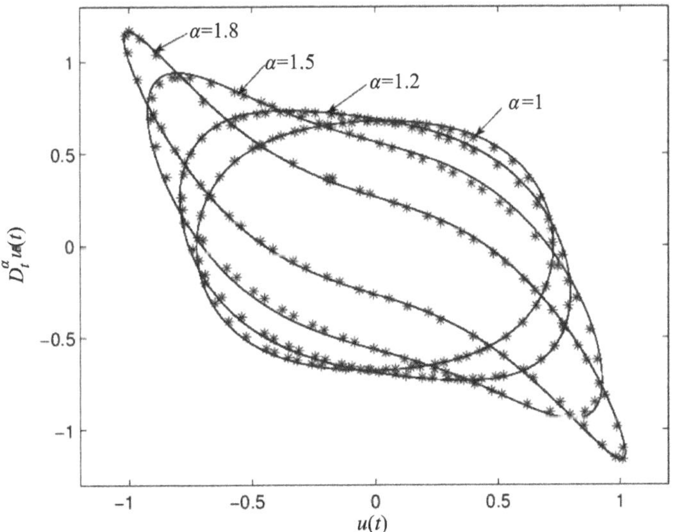

Figure 3.4b Phase portraits for fractional order $\alpha = \{1,1.2,1.5,1.8\}$.

Solid line: analytical solution, Stars: numerical result

3.1.5 Concluding Remarks

In this section, we have extended the method of harmonic balance in combination with homotopy continuation to investigate a Duffing oscillator with fractional order derivative. The fractional derivative is described in the Caputo sense. The relations of fractional order versus steady state amplitude, phase angle and frequency versus steady state amplitude are determined by finding all the steady state solutions. The influence of fractional derivative representations on the steady state response and some nonlinear phenomena such as superharmonic resonance, supercritical pitchfork bifurcation, symmetry breaking, hysteresis and jump of Duffing oscillator with damping defined by fractional derivative have been examined and investigated analytically. As we can see that, the method of harmonic balance in combination with polynomial homotopy continuation has great potential to be applied to fractionally damped systems. In the next sections, this method shall be generalized to study the other nonlinear systems with fractional order derivative.

3.2 Transition Curves and Bifurcations of a Class of Fractional Mathieu–type Equations

3.2.1 Introduction

The well-known Mathieu equation, first introduced by E. Mathieu when he studied vibrating elliptical membranes [E. Mathieu,1868], is a linear differential equation with periodic coefficients. Its canonical form is

$$u'' + (k_1 + k_2 \cos t)u = 0 \tag{3.8}$$

The equation, which commonly occurs in nonlinear vibration problems, is useful in many mathematics and physics problems. In recent years, the study of oscillatory behaviors in fractional order dynamical systems is also a subject of increasing attention [I. Podlubny, 1999; Y.A. Rossikhin & M.V. Shitikova, 2010]. It has been found that the fractional system can generate regular oscillation like integer order system. Therefore, it may be of interest to consider some modifications to the well-known equations, which the dependent variable and/or its derivatives occur to be fractional power. For instance, fractional linear oscillator [M. Naber, 2010; Y.G. Kang & X. Zhang, 2010], fractional Duffing equation [P. Arena et al., 2000; P. Wahi & A. Chatterjee, 2004], fractional van der Pol equation [M. Barbosa et al., 2007; Z. Guo et al., 2011; F. Xie & X. Lin, 2009] and fractional Jerk equation [W.M. Ahmad & J.C. Sprott, 2003] have been presented and applied. For the Mathieu type equation, the chaotic behaviors of a nonlinear damped Mathieu system was investigated by applying numerical integration method [Z.M. Ge & C.X. Yi, 2007]. The transition curves separating the regions of stability in the following fractional order Mathieu equation have been recently studied [R.H. Rand et al., 2010]

$$u'' + hD_t^\alpha u + (k_1 + k_2 \cos t)u = 0 \tag{3.9}$$

In this section, we shall consider a slightly different version of the fractional Mathieu equation by introducing a fractional order time derivative of the damped Mathieu equation,

$$D_t^{1+\alpha} u + hD_t^\alpha u + (k_1 + k_2 \cos \omega t)u = 0 \tag{3.10}$$

where $0 < \alpha \leq 1$; is the linear damping rate; k_1 is the linear spring constant; k_2 and ω are the driving amplitude and the driving frequency, often called the pump frequency. A similar approach was performed for the Duffing [P. Arena et al., 2000], van der Pol [R.S. Barbosa et al., 2007; Z. Guo et al., 2011] and Chua [T.T. Hartley et al., 1995] equations.

A more realistic model can be developed by including some nonlinear terms in the Mathieu equation such as cubic nonlinear stiffness. In this section, we shall also investigate a general version of fractional Mathieu-Duffing equation of the form

$$D_t^{1+\alpha} u + hD_t^\alpha u + (k_1 + k_2 \cos \omega t)u + k_3 u^3 = 0 \tag{3.11}$$

where k_3 is the nonlinear stiffness constant, or the Duffing parameter. Note that the systems (3.10) and (3.11) obviously have trivial solution $u=1$. In this section, we use the method of harmonic balance to obtain approximate expressions of the stability tongue regions for the fractional Mathieu equation (3.10). Then, the steady state response of Eq. (3.11) is investigated by applying the technique of residue harmonic balance in combination with homotopy continuation. In addition, some nonlinear phenomena such as symmetric breaking and period doubling bifurcation are exhibited analytically.

3.2.2 Transition Curves of a General Version of Fractional Mathieu Equation

In this section, we use the harmonic balance to obtain the approximate expressions for the transition curves in the fractional Mathieu equation (3.10). Based on the perturbation theory and analysis from [R.H. Rand, 2010], it may be supposed that on the transition curves, there exist periodic solution to Eq. (3.10) with period $2\pi/\omega$ or $4\pi/\omega$. In order to obtain an approximation with period $4\pi/\omega$ for $n=1$, we assume a truncated Fourier series:

$$u = a_1 \cos\frac{\omega t}{2} + b_1 \sin\frac{\omega t}{2} \tag{3.12}$$

Substituting Eq. (3.12) into Eq. (3.10), collecting terms and equating the coefficients of $\cos\dfrac{\omega t}{2}$ and $\sin\dfrac{\omega t}{2}$ to zero, we obtain two equations as below:

$$(\frac{\omega}{2})^{1+\alpha}(-a_1\sin\frac{\alpha\pi}{2}+b_1\cos\frac{\alpha\pi}{2})+k_1a_1+\frac{1}{2}k_2a_1+h(\frac{\omega}{2})^{\alpha}(a_1\cos\frac{\alpha\pi}{2}+b_1\sin\frac{\alpha\pi}{2})=0$$

$$(\frac{\omega}{2})^{1+\alpha}(-b_1\sin\frac{\alpha\pi}{2}-a_1\cos\frac{\alpha\pi}{2})+k_1b_1-\frac{1}{2}k_2b_1+h(\frac{\omega}{2})^{\alpha}(-a_1\sin\frac{\alpha\pi}{2}+b_1\cos\frac{\alpha\pi}{2})=0$$

$$\tag{3.13}$$

Eliminating a_1 and b_1 from Eq. (3.13) gives the following approximate expression for the transition curves:

$$k_1 = \left(\frac{\omega}{2}\right)^{1+\alpha} \sin\frac{\alpha\pi}{2} - h\left(\frac{\omega}{2}\right)^{\alpha} \cos\frac{\alpha\pi}{2} \pm \sqrt{\frac{1}{4}k_2^2 - [\left(\frac{\omega}{2}\right)^{1+\alpha} \cos\frac{\alpha\pi}{2} + h\left(\frac{\omega}{2}\right)^{\alpha} \sin\frac{\alpha\pi}{2}]^2}$$

$$(3.14)$$

In a similar fashion we may obtain approximations with period $2\pi/\omega$ for the other transition curves for the $n=0$ tongue. For instance, we assume a truncated Fourier series:

$$u = a_0 + a_1\cos(\omega t) + b_1\sin(\omega t) \tag{3.15}$$

Substituting Eq. (3.15) into Eq. (3.10), collecting terms and equating the coefficients of $\cos(\omega t)$, $\sin(\omega t)$ and constant term to zero, we obtain the following three equations

$$\omega^{1+\alpha}(b_1\cos\frac{\alpha\pi}{2} - a_1\sin\frac{\alpha\pi}{2}) + k_1 a_1 + k_2 a_0 + h\omega^{\alpha}(b_1\sin\frac{\alpha\pi}{2} + a_1\cos\frac{\alpha\pi}{2}) = 0$$

$$\omega^{1+\alpha}(-b_1\sin\frac{\alpha\pi}{2} - a_1\cos\frac{\alpha\pi}{2}) + k_1 b_1 + h\omega^{\alpha}(b_1\cos\frac{\alpha\pi}{2} - a_1\sin\frac{\alpha\pi}{2}) = 0 \tag{3.16}$$

$$k_1 a_0 + \frac{1}{2}k_2 a_1 = 0$$

Eliminating a_0, a_1 and b_1 from Eq. (3.16) gives the following approximate expression for the transition curves:

$$2k_1[(\omega^{1+\alpha}\cos\frac{\alpha\pi}{2} + h\omega^{\alpha}\sin\frac{\alpha\pi}{2})^2 + (\omega^{1+\alpha}\sin\frac{\alpha\pi}{2} - k_1 - h\omega^{\alpha}\cos\frac{\alpha\pi}{2})^2]$$

$$+k_2^2(\omega^{1+\alpha}\sin\frac{\alpha\pi}{2} - k_1 - h\omega^{\alpha}\cos\frac{\alpha\pi}{2}) = 0 \tag{3.17}$$

Figures 3.5 and 3.6 show the transition curves (3.14) in the case of $n=1$ tongue for $\omega=1$ and $\omega=2$ respectively. From Figs. 3.5 and 3.6, it is observed that a change in the fractional order derivative α affects the shape and location of the transition curves. The instability region shrinks with higher α and higher ω. As the fractional order α decreases, the location of the driving amplitude k_2 on transition curves increases for $\omega=1$ and $\omega=2$. The shape of the transition curves of the fractional Mathieu Eq. (3.10) is different from the corresponding transition curves of the fraction Mathieu equation (3.9). Figs. 3.7 and 3.8 show the transition curves

(3.17) for the stability tongue at $n=0$ for $\omega=1$ and $\omega=2$, respectively. We note that the shape of the transition curve does not change much for fractional order $\alpha \in [0.5,1]$.

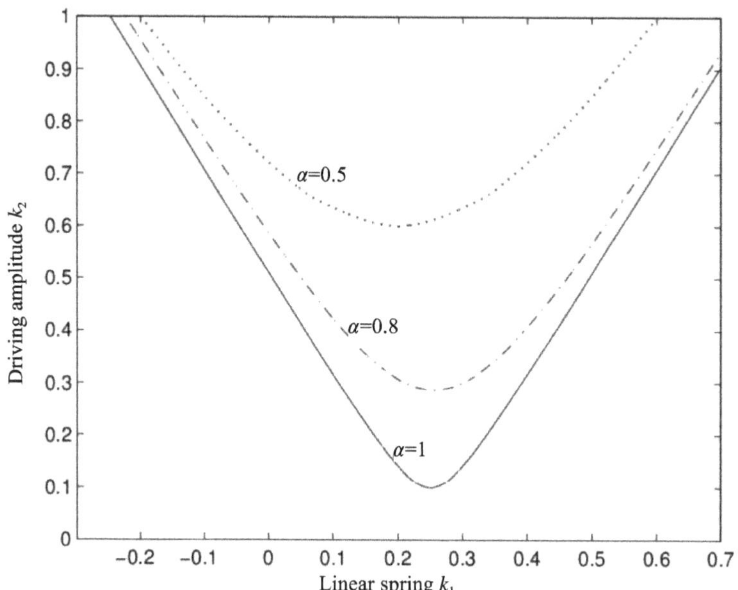

Figure 3.5 Stability tongue in the fractional Mathieu equation (3.10) for h=0.1, ω=1, n=1 and $\alpha=0.5,0.8,1$.

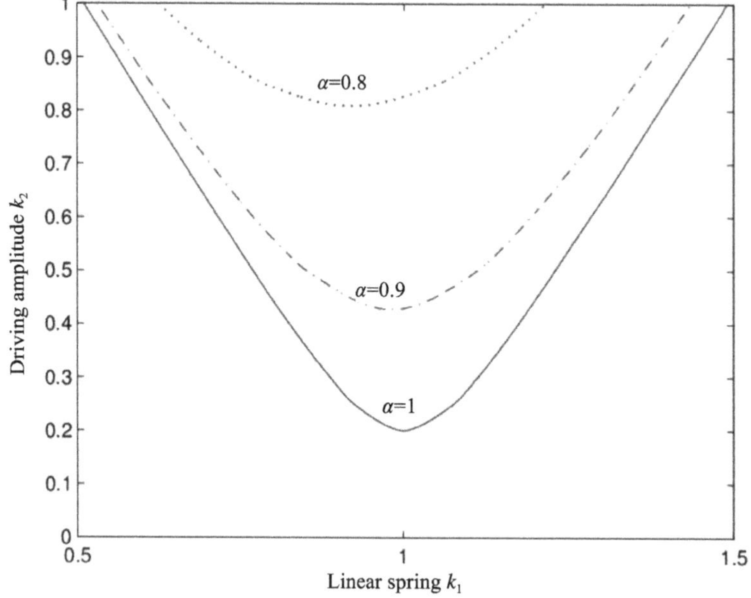

Figure 3.6 Stability tongue in the fractional Mathieu equation (3.10) for h=0.1, ω=2, n=1 and $\alpha=0.8,0.9,1$.

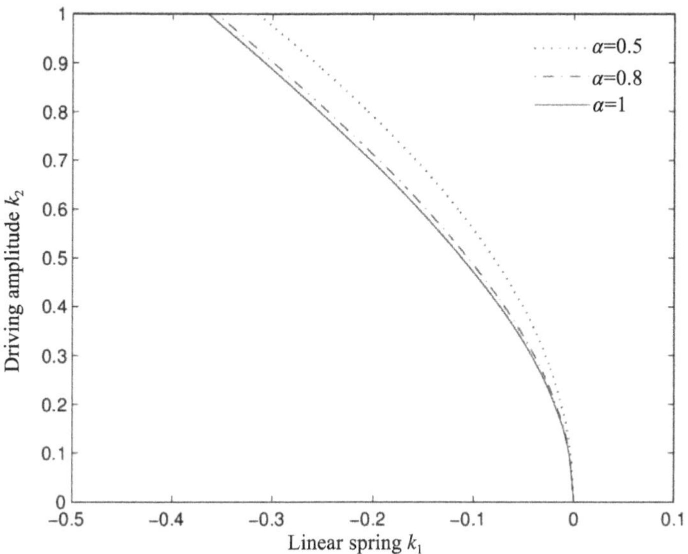

Figure 3.7 Stability tongue in the fractional Mathieu equation (3.10)
for h=0.1, ω=2, n=0 and α = 0.5,0.8,1.

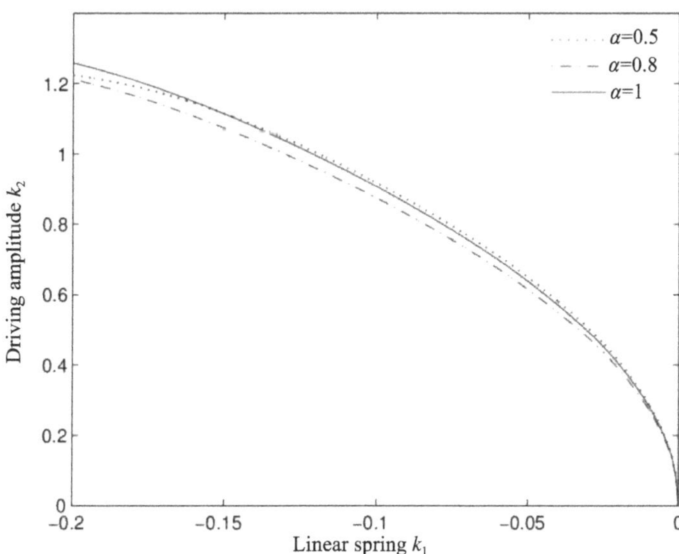

Figure 3.8 Stability tongue in the fractional Mathieu equation (3.10)
for h=0.1, ω=2, n=0 and α = 0.5,0.8,1.

3.2.3 Steady State Response of the Fractional Order Mathieu-Duffing Equation

Based on the fractional Mathieu equation, we introduce the cubic nonlinear stiffness and investigate a more general version of fractional Mathieu-Duffing equation

$$D_t^{1+\alpha}u - (k_1 + k_2 \cos \omega t)u + hD_t^\alpha u + k_3 u^3 = 0 \tag{3.18}$$

in this section.

For the periodic solution to the fractional Mathieu equation (3.10), we assume that the steady state response of Eq. (3.18) can be expanded in Fourier series as follows:

$$u(t) = a_0 + a_1 \cos \frac{\omega t}{2} + b_1 \sin \frac{\omega t}{2} + a_2 \cos(\omega t) + b_2 \sin(\omega t) + a_3 \cos \frac{3\omega t}{2} + b_3 \sin \frac{3\omega t}{2} \tag{3.19}$$

For convenience, we introduce a new time scale $\tau = \omega t/2$ of period 2π to transform Eqs. (3.18) and (3.19) to

$$(\frac{\omega}{2})^{1+\alpha} D_\tau^{1+\alpha}u - (k_1 + k_2 \cos 2\tau)u + h(\frac{\omega}{2})^\alpha D_\tau^\alpha u + k_3 u^3 = 0 \tag{3.20}$$

$$u(\tau) = a_0 + a_1 \cos \tau + b_1 \sin \tau + a_2 \cos 2\tau + b_2 \sin 2\tau + a_3 \cos 3\tau + b_3 \sin 3\tau \tag{3.21}$$

Substitute Eq. (3.21) into Eq. (3.20) to obtain the following harmonic balance equations

$$\begin{cases} R_0(a_0,a_1,a_2,a_3,b_1,b_2,b_3) = \dfrac{2}{\pi} \displaystyle\int_0^\pi \Phi(D_\tau^{1+\alpha}u, D_\tau^\alpha u, u, \lambda, \tau)\,d\tau = 0, \\[2mm] R_i^c(a_0,a_1,a_2,a_3,b_1,b_2,b_3) = \dfrac{2}{\pi} \displaystyle\int_0^\pi \Phi(D_\tau^{1+\alpha}u, D_\tau^\alpha u, u, \lambda, \tau)\cos i\tau\,d\tau = 0, i=1,2,3, \\[2mm] R_i^s(a_0,a_1,a_2,a_3,b_1,b_2,b_3) = \dfrac{2}{\pi} \displaystyle\int_0^\pi \Phi(D_\tau^{1+\alpha}u, D_\tau^\alpha u, u, \lambda, \tau)\sin i\tau\,d\tau = 0, i=1,2,3. \end{cases} \tag{3.22}$$

Carrying out the integration yields the nonlinear algebraic equations for the determination of the coefficients

$$[K(q,\lambda)]\{q\} = \{R\} = 0 \tag{3.23}$$

where $\{R\}=[R_0, R_1^c, R_1^s, R_2^c, R_2^s, R_3^c, R_3^s]^T$ is the vector of residues to be annihilated, $\{q\} = [2a_0, a_1, b_1, a_2, b_2, a_3, b_3]^T$ is the Fourier coefficient vector of the response and $[K(q,\lambda)]$ is the total stiffness matrix. Using the sum and product formulae for the trigonometric functions, the expressions of harmonic balance equations are given.

For the steady state approximate solution (3.19) of Eq. (3.18), we find that:

(i) When $u(t)$ is a solution of Eq. (3.20), then $-u(t)$, $u(t+2\pi/\omega)$ and $-u(t+2\pi/\omega)$ are also solutions. I.e. the steady state harmonic balance solution (3.19) is invariant under the transformation

$$(a_0,a_1,b_1,a_2,b_2,a_3,b_3) \mapsto (a_0,-a_1,-b_1,a_2,b_2,-a_3,-b_3)$$
$$\mapsto -(a_0,a_1,b_1,a_2,b_2,a_3,b_3) \mapsto -(a_0,-a_1,-b_1,a_2,b_2,-a_3,-b_3)\;.$$

(ii) When the bias term A_0 is zero, the response (3.19) is symmetric and that denoted as S_1

$$u(t) = a_1 \cos\frac{\omega t}{2} + b_1 \sin\frac{\omega t}{2} + a_3 \cos\frac{3\omega t}{2} + b_3 \sin\frac{3\omega t}{2}$$

(iii) When A_1 and A_3 are zero simultaneously, the response (3.19) becomes $u(t) = a_0 + a_2 \cos\omega t + b_2 \sin\omega t$ which is denoted as S_2.

(iv) Otherwise, all the A_0, A_1, A_2, A_3 are nonzero, we denote the response (3.19) as S_3, where $A_0 = |a_0|, A_1 = \sqrt{a_1^2 + b_1^2}, A_2 = \sqrt{a_2^2 + b_2^2}, A_3 = \sqrt{a_3^2 + b_3^2}$

Furthermore, we denote the maximal steady state amplitude $A_{\max}$ as

$$A_{\max} = \max_{\forall t}[a_0 + a_1 \cos\frac{\omega t}{2} + b_1 \sin\frac{\omega t}{2} + a_2 \cos(\omega t) + b_2 \sin(\omega t) + a_3 \cos\frac{3\omega t}{2} + b_3 \sin\frac{3\omega t}{2}]$$

for presenting results later.

3.2.4 Numerical Examples

By means of technique of residue harmonic balance in combination with polynomial homotopy continuation which described in chapter 1, all the isolated solutions and the corresponding improved corrections of the harmonic balance equations (3.23) can be obtained. For definite comparison, the parameter values $\omega = 2$, $h = 0.25$, $k_1 = 1$, $k_3 = 1$ are selected. In all figures shown in this section, solid lines denote stable branch, broken lines denote unstable branch and dots is result from numerical integration. The predicted limit cycles are stable which can be observed in the numerical simulation with the steady state solutions as initial conditions, and consequently they will not be observed for unstable solutions in the numerical simulation.

3.2.4.1 Fractional order versus amplitude

In this subsection, the steady state responses are computed for the fixed driving amplitude $k_2 = 4.6$. The steady state approximate solution is assumed to be of the form of Eq. (3.19). Figs. 3.9 and 3.10 show the relations of fractional order versus response amplitude for A_0 and A_{max} respectively. From the Figs. 3.9 and 3.10, it is observed that when the fractional order is smaller, the steady state response is of symmetry form S_1. As the fractional order increases to 0.76708, the symmetric solution S_1 branches out to an unstable asymmetric solution S_3, while the original stable solution has a remarkable change in the slope of the amplitude A_{max}. The asymmetric solution S_3 and symmetric solution S_1 change stabilities at the same time with the increasing of the fractional order α to 0.88484. When the fractional order increases further to 0.985046, a pair of new branch solutions S_2 occur, in which the upper branch is stable and other is unstable.

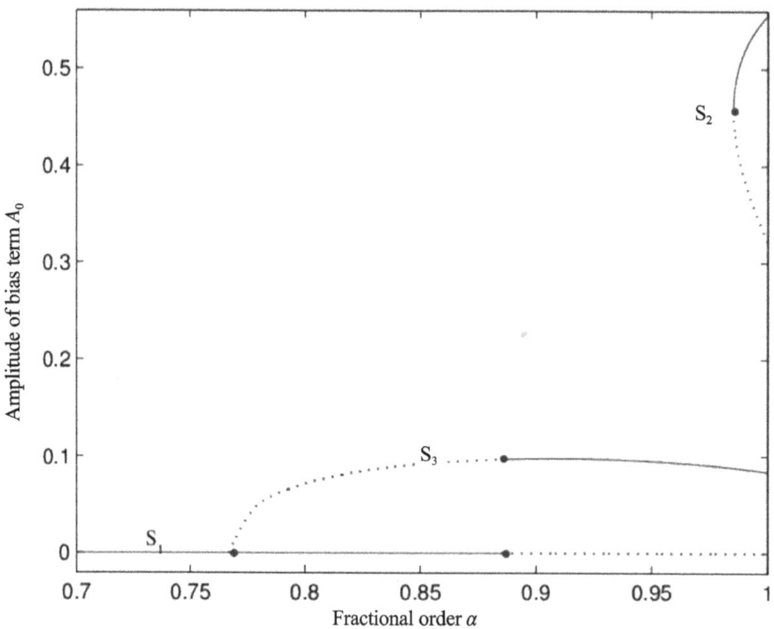

Figure 3. 9 Fractional order versus response amplitude A_0 curves

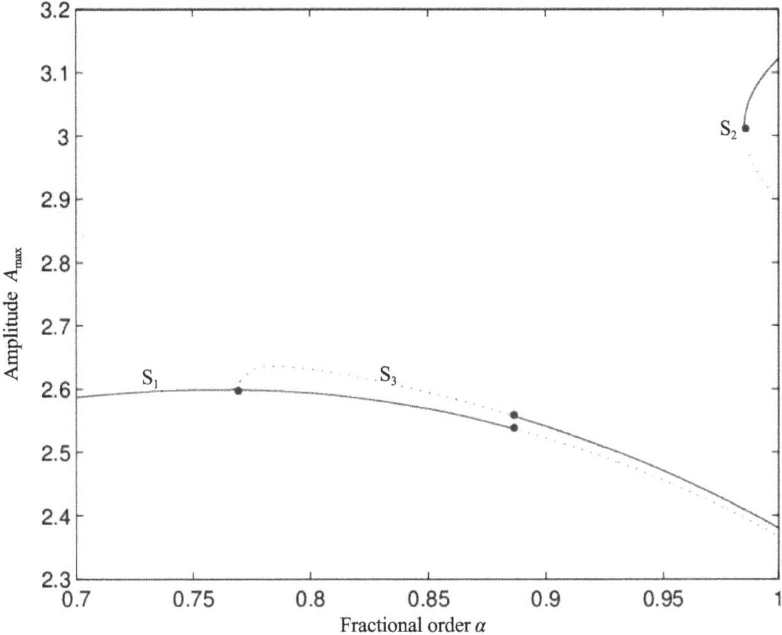

Figure 3. 10 Fractional order versus response amplitude $A_{\max}$ curves

To verify, we compare the obtained stable solutions and numerical integration results for fractional order $\alpha = 0.5, 0.8, 0.95$ and 1. The analytical expressions of symmetric solutions for $\alpha = 0.5, 0.08$ and the asymmetric solutions for $\alpha = 0.95, 1$ are tabulated in Table 3.2, respectively. When two or more stable solutions coexist, the initial conditions determine which steady state solution prevails. It can be seen that the higher-order superharmonic coefficients are in general very small comparing to their lower harmonic counterparts. Figs. 3.11–3.14 show the phase portraits of corresponding solutions for $\alpha = 0.5, 0.8, 0.95$ and $\alpha = 1$ respectively, it is found that the analytical solutions match well with the numerical integration solutions. Two asymmetrical solutions with forms S_2 and S_3 coexist when $\alpha = 1$. Therefore, the initial conditions $u(0) = 2.5$, $u'(0) = 4$ and $u(0) = 1$, $u'(0) = 0.1$ are selected in Figs. 3.14a and 3.14b for numerical integration calculation, respectively.

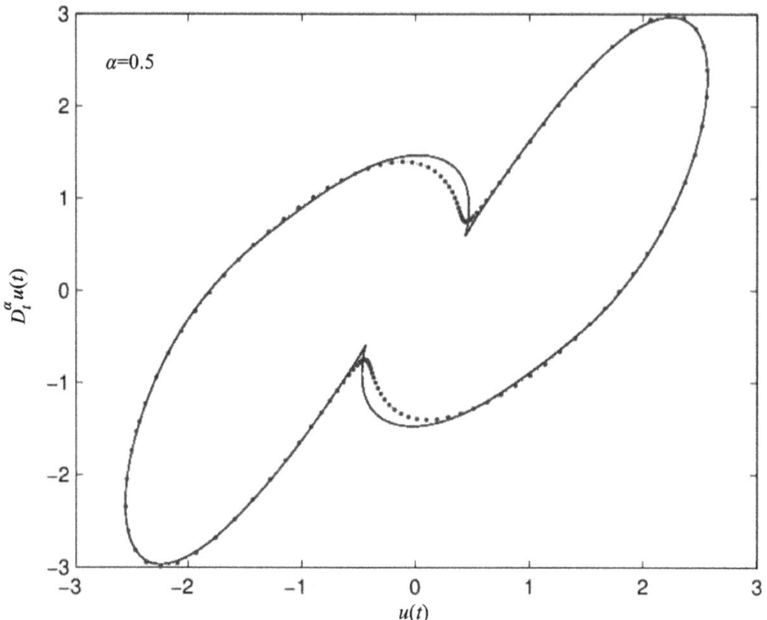

Figure 3.11 Phase portrait for fractional order $\alpha = 0.5$

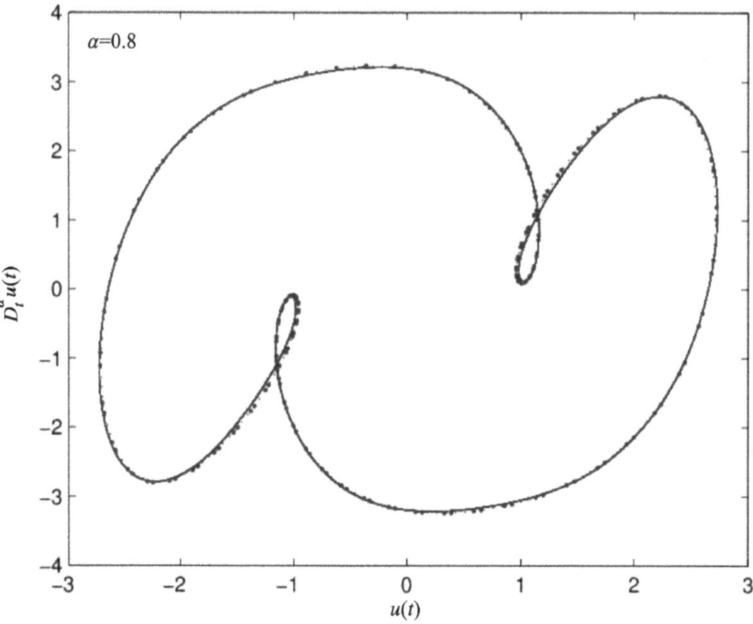

Figure 3.12 Phase portrait for fractional order $\alpha = 0.8$

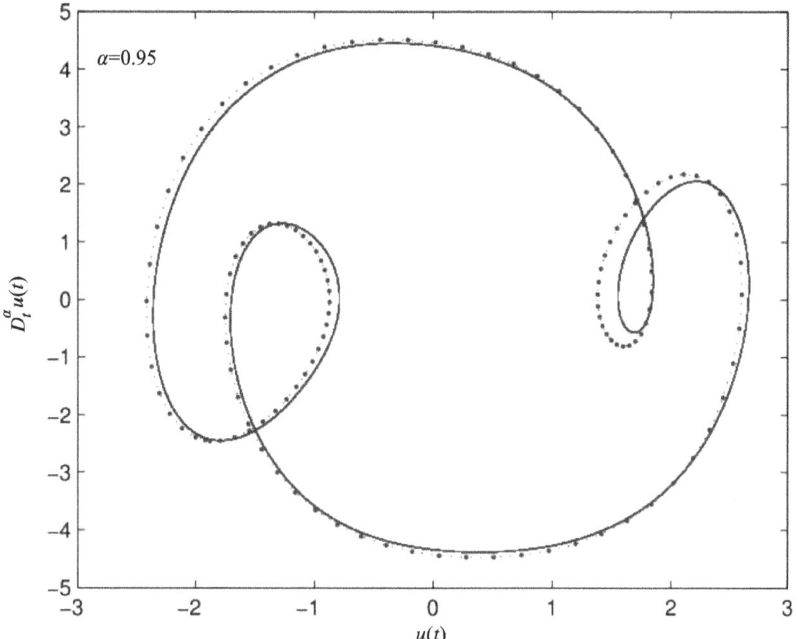

Figure 3.13 Phase portrait for fractional order $\alpha = 0.95$

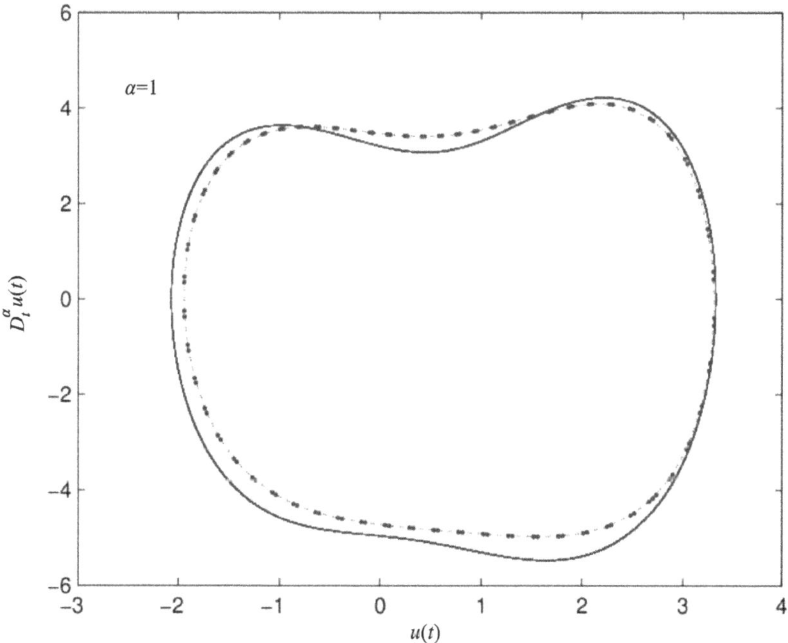

Figure 3.14a Phase portrait of solution with form S$_2$ for $\alpha = 1$

The initial condition $u(0) = 2.5,\ u'(0) = 4$ **is applied for numerical simulation.**

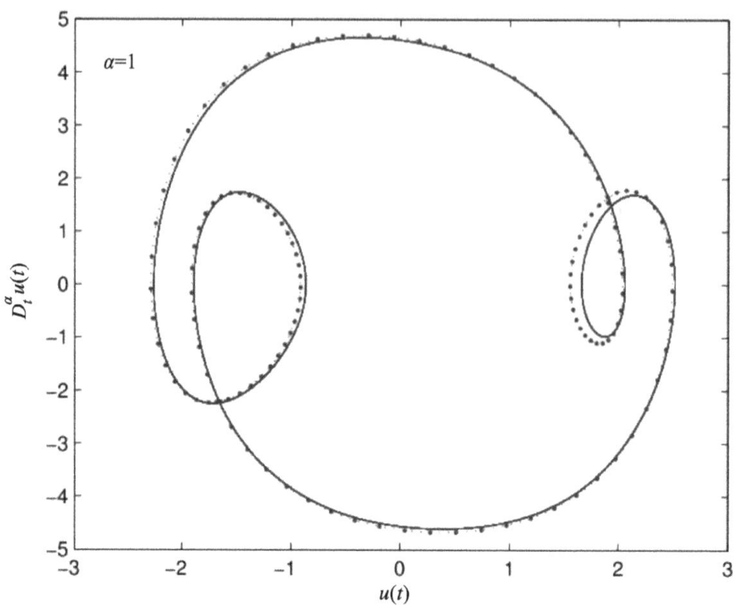

Figure 3.14b Phase portrait of solution with form S_3 for $\alpha = 1$. The initial condition $u(0) = 1$, $u'(0) = 0.1$ is applied for numerical simulation

Table 3.2 Analytical expressions of steady state solution for Eq. (3.18) under $\alpha = 0.5, 0.8, 0.95$ and 1

	$\alpha = 0.5$	$\alpha = 0.8$	$\alpha = 0.95$	$\alpha = 1$	
Constant			0.08233	0.53770	0.07682
$\cos(\omega t/2)$	1.70629	2.02166	2.05339		2.10172
$\sin(\omega t/2)$	0.43333	0.06386	0.00756		0.01578
$\cos(\omega t)$			0.46165	2.29416	0.44004
$\sin(\omega t)$			-0.09985	0.94079	-0.04729
$\cos(3\omega t/2)$	0.10738	-0.18108	-0.69592		-0.81045
$\sin(3\omega t/2)$	0.65117	0.78357	0.4917		0.26809
$\cos(2\omega t)$			-0.04714	-0.16635	-0.07734
$\sin(2\omega t)$			0.11839	0.28604	0.06026
$\cos(5\omega t/2)$	0.02241	-0.07396	-0.0089		0.00463
$\sin(5\omega t/2)$	0.13871	0.06897	-0.02586		-0.02199
$\cos(3\omega t)$			-0.02861	0.03721	0.03262
$\sin(3\omega t)$			0.02806	0.14364	0.01037
$\cos(7\omega t/2)$	-0.02728	-0.04429	0.00419		0.01620

continued

	$\alpha=0.5$	$\alpha=0.8$	$\alpha=0.95$	$\alpha=1$	
$\sin(7\omega t/2)$	0.04079	-0.00516	-0.01904		-0.01236
$\cos(4\omega t)$			-0.00188	-0.03462	0.00615
$\sin(4\omega t)$			-0.01234	0.00628	-0.00775
$\cos(9\omega t/2)$	-0.01334	-0.00372	0.00072		0.00127
$\sin(9\omega t/2)$	0.00879	-0.00915	0.00236		0.00221

3.2.4.2 Bifurcations under different fractional orders

In this subsection, we study the variation of the driving amplitude k_2 in the interval from 4.2 to 5.2 and take the fractional order $\alpha = 0.8, 0.9, 0.95, 0.99$ and 1, respectively. The steady state solutions (3.19) can be obtained by solving Eq. (3.23). The influence of the fractional order is examined by investigating the relations of the driving amplitude versus the response amplitudes. Some nonlinear phenomena, such as symmetric breaking, saddle-node bifurcation and period doubling bifurcation are illustrated analytically.

Figures. 3.15–3.19 illustrate relation of the driving amplitude k2 versus response amplitudes for fractional order $\alpha = 0.8, 0.9, 0.95, 0.99$ and 1, respectively. From Figs. 3.15–3.17, it can be observed that there are two branch solutions when increasing the driving amplitude k_2 from 4.2 to 5.2 for all cases of fractional order $\alpha = 0.8, 0.9, 0.95$. However, the stability of branch solutions is quite different. When we take fractional order $\alpha = 0.8$, the symmetric solution with form S_1 keeps stable all the time and stable symmetric solution S_1 bifurcates an unstable asymmetric solution with form S_3 when driving amplitude increases to 4.5147. When fractional order $\alpha = 0.9$, the stable symmetric branch solution loses stability and bifurcates a stable asymmetric branch solution with the driving amplitude k_2 increasing to 4.2685. Further, when fractional order $\alpha = 0.95$, the asymmetric branch solution occurs at point $k_2 \approx 4.219$, and the stable asymmetric branch solution loses stability at point $k_2 \approx 4.845$. From Figs. 3.15–3.17, we can find that the shape of steady state amplitude is almost same, but the bifurcation point of branch solutions is completely different for the different fractional order systems with $\alpha = 0.8, 0.9, 0.95$.

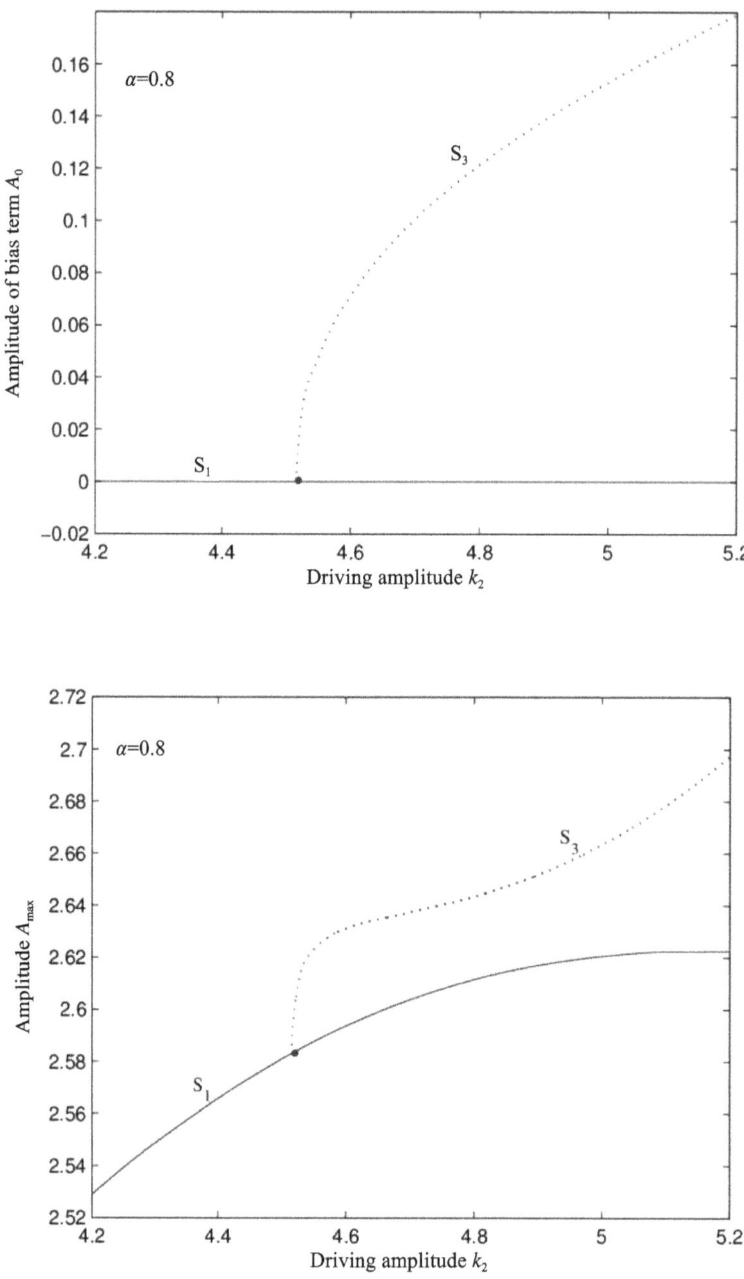

Figure 3.15 Driving amplitude k_2 versus response amplitudes for $\alpha = 0.8$

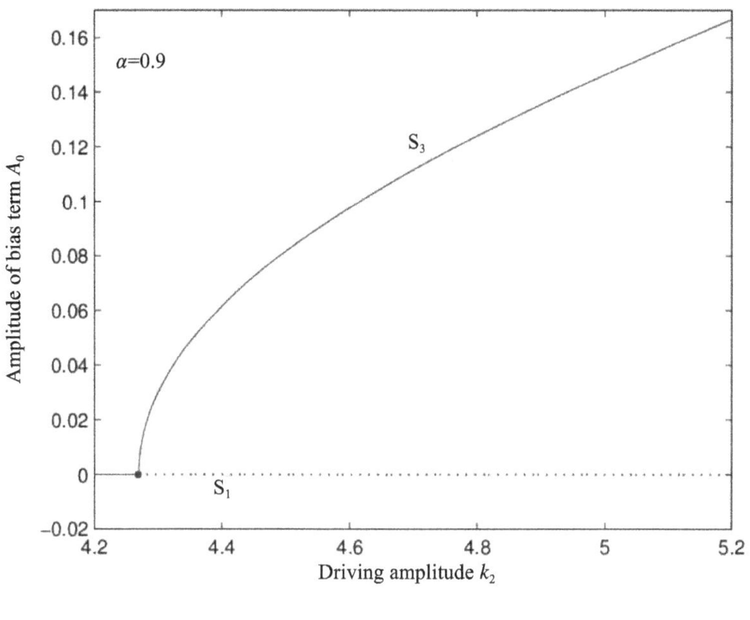

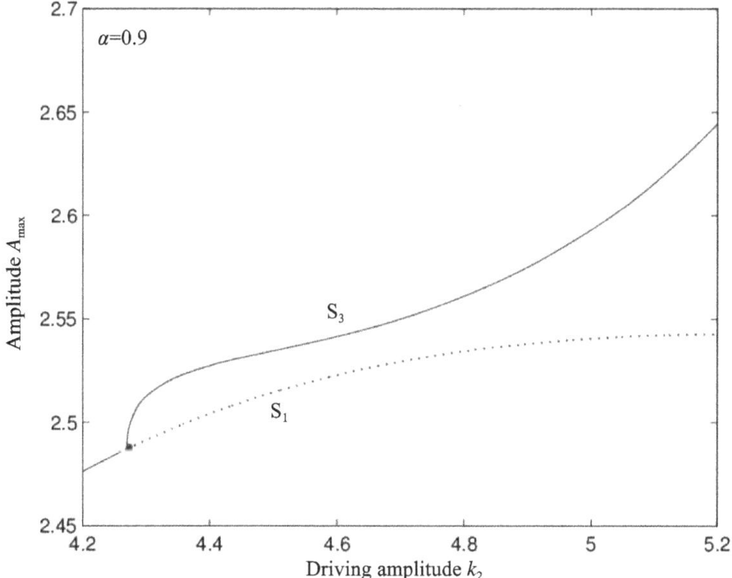

Figure 3.16 Driving amplitude k_2 versus response amplitudes for $\alpha = 0.9$

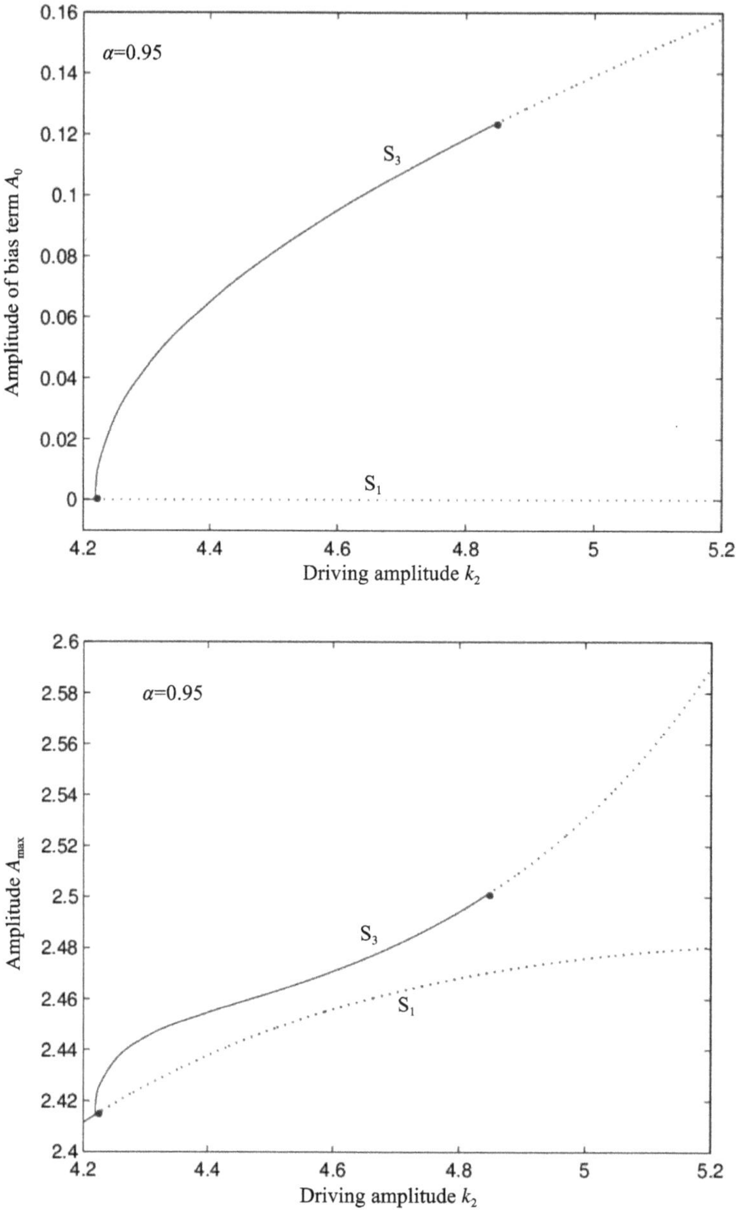

Figure 3.17 Driving amplitude k_2 versus response amplitudes for $\alpha = 0.95$

Figures 3.18 and 3.19 show the driving amplitude versus response amplitude curves for $\alpha = 0.99$ and 1. It should be noted that the system (3.18) reduces to an integer order system when $\alpha = 1$. In general, the shape, number and stability of the branch solutions are almost same for system with $\alpha = 0.99$ and 1, which are shown in Figures 3.18–3.19. That is, the steady state response of fractional order system with $\alpha = 0.9$ converges to integer order system. In more specific terms, it is observed for the fractional system with $\alpha = 0.99$ that:

(i) Symmetric breaking phenomena emerge when the driving amplitude k_2 increases to 4.2485 where the stable symmetric solution S_1 loses stability and bifurcates to a stable asymmetric solution S_3. After that, there are four periodic solutions, two are stable and the other two are unstable.

(ii) When the driving amplitude k_2 increases from 4.2 to 4.49331, a saddle-node bifurcation occurs, that is, a pair of fundamental asymmetric solution S_2 occurs.

(iii) Period doubling bifurcation occurs when the driving amplitude k_2 increases to 4.712, the stable fundamental solution S_2-form loses stability and bifurcates to a stable period-2 solution. Meanwhile, another asymmetric period-2 solution S_3 loses stability.

Similarly, the above corresponding bifurcation occur at point $k_2 \approx 4.26405$ (saddle-node bifurcation), $k_2 \approx 4.2685$ (Symmetric breaking), $k_2 \approx 4.666$ (Period doubling bifurcation) for integer order system with $\alpha = 1$.

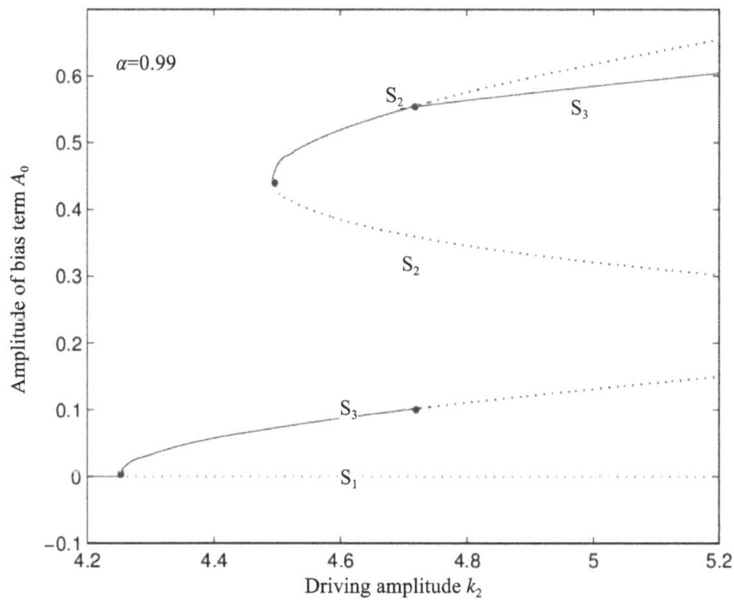

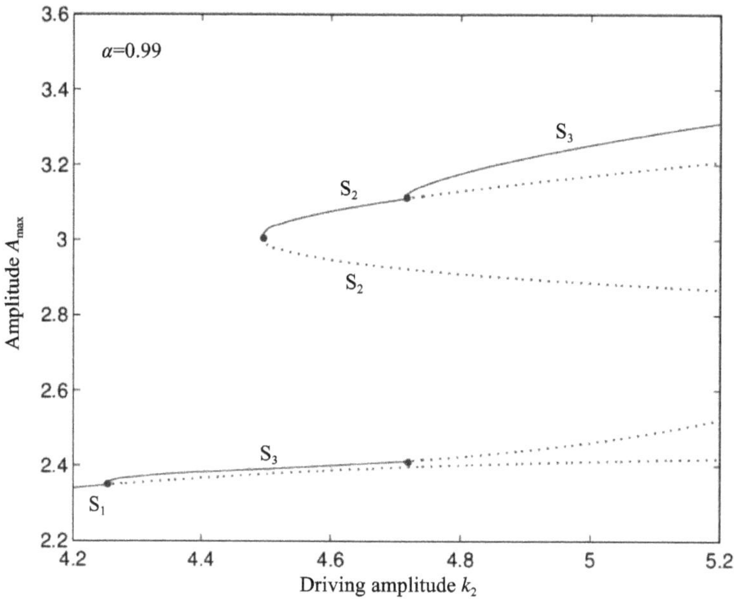

Figure 3.18a Driving amplitude k_2 versus bias term amplitude for $\alpha = 0.99$

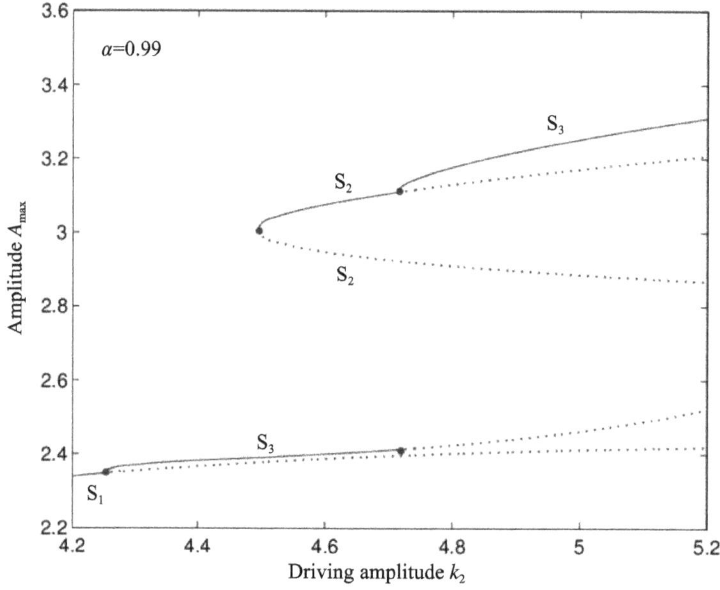

Figure 3.18b Driving amplitude k_2 versus maximal response amplitude for $\alpha = 0.99$

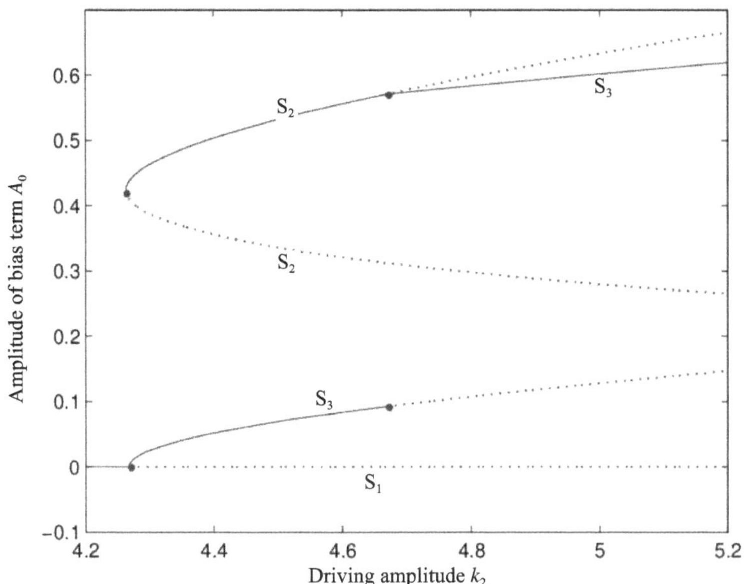

Figure 3.19a Driving amplitude k_2 versus bias term amplitude for $\alpha = 1$

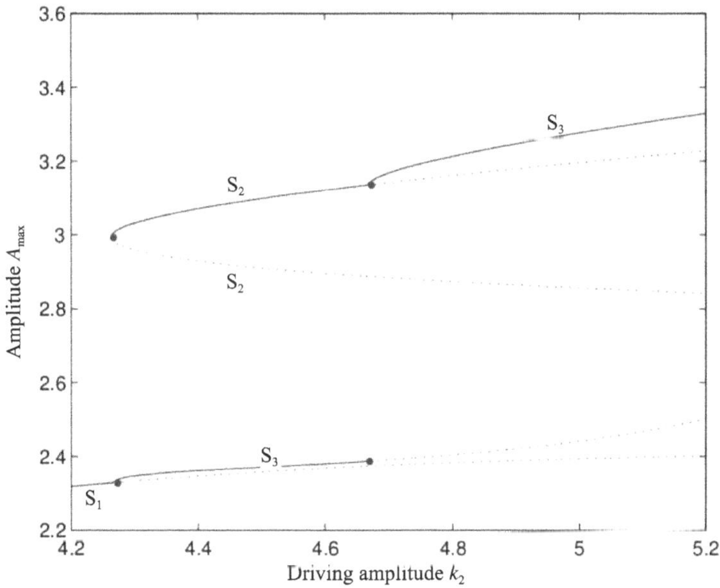

Figure 3.19b Driving amplitude k_2 versus maximal response amplitude for $\alpha = 1$

3.2.5 Concluding Remarks

The approximate expressions for the tongue diagram corresponding to $n = 1$ and $n = 0$ transition curves separating regions of stability from regions of instability in the fractional Mathieu equation have been obtained via the method of harmonic balance. We showed that a bigger change of the shape and location of the $n=1$ transition curve by changing the value of the order of the fractional derivative and the driving frequency. However, the shape of the transition curve does not change very much for different fractional orders in the case of the $n=0$ tongue.

Based on the fractional Mathieu equation, a corresponding fractional Mathieu-Duffing equation has been established and the steady state approximate response are analyzed analytically by means of the technique of residue harmonic balance in combination with polynomial homotopy continuation. The curves with respect to fractional order versus response amplitude, parameter driving amplitude versus response amplitude with different fractional orders have been illustrated analytically. It can be found that the bifurcation point and stability of the branch solutions are different under different fractional systems. When the fractional order increases to some certain values, rich nonlinear phenomena such as the symmetric breaking, saddle-node bifurcation as well as period-doubling bifurcation phenomena have been found and exhibited analytically at the end.

Chapter 4 Resonance Response and Effects of Fractional Differentiation Orders on Coupled Vibration Systems

4.1 Effects of Fractional Differentiation Orders on a Coupled Duffing Circuit

4.1.1 Introduction

Over the last few years, it has been proved that the behavior of many physical systems can be properly described by the fractional differential equation because the fractional derivatives provide an excellent instrument for the description of memory and hereditary properties of various materials and processes [I. Podlubny, 1999; R.L. Bagley & P.J. Torvik, 1983, 1985]. For instance, Bagley et al [R.L. Bagley & P.J. Torvik, 1983, 1985] have shown that fractional derivative models describe very well the frequency-dependent damping behavior of materials and systems in the earlier studies.

Recently, fractional differential equations appear more and more frequently in electronic engineering [D. Cafagna & G. Grassi, 2008; P. Ivo, 2010]. Classical electrical circuits consist of resistors and capacitors and are described by integer order models. However, circuits may have the so-called fractance, which represents an electrical element with fractional order impedance. To put it more specifically, Pu et al [Y.F. Pu et al., 2005] put forward a 1/2 order net-grid-type analog fractance circuit and they found that the fractance circuit can be found by common circuit elements such as resistance, capacitance or inductance by some high self-similar fracture. Westerlund and Ekstam [S. Westerlund & L. Ekstam, 1994] proposed a new linear capacitor model containing fractional derivative, they provided the table of various capacitor dielectrics with appropriated fractional order which has experimentally been obtained by measurements. In [S. Westerlund, 2002], Westerlund described the behavior of a real inductor, for a general current in the inductor, the voltage can be described by a fractional order model. Pereira et al [E. Pereira et al., 2004] substituted the capacitance by a fractance in the nonlinear RLC circuit to obtain a fractional version of the van der Pol equation. In [Y. Ma & H. Kawakami, 2002, 2003], Ma and Kawakami defined a coupled Duffing circuit, which is a set of four ordinary differential equations. Moreover, they studied the fundamental combinatorial nonlinear resonances of the system and obtained the bifurcation diagram under multiple cases of external periodic force. Garcia [N. Garcia, 2010] presented a time-domain method to determine the periodic steady state solution of the above equation of coupled Duffing circuit.

In this section, we extend the above classical coupled Duffing circuit model to fractional order system, and predict the fundamental resonance response of fractional order coupled Duffing circuit system in detail using the method of residue harmonic balance in combination with homotopy continuation. Besides, the influences of different commensurate and incommensurate fractional orders on the frequency, excitation amplitude and fractional order versus response curves are discussed.

4.1.2 Fractional Order Model of a Coupled Duffing Circuit

There are a large number of electric and magnetic phenomena involve the application of fractional calculus. Based on Curie's empirical law, Westerlund et al [S. Westerlund, 2002] proposed a new linear capacitor model. For a general input voltage, the current is

$$I(t) = C\frac{\mathrm{d}^{\alpha}V(t)}{\mathrm{d}t^{\alpha}} \equiv C_0 D_t^{\alpha}V(t) \tag{4.1}$$

where C is the capacitance of the capacitor which related to a kind of dielectric. α is a fractional order which related to the losses of the capacitor. Westerlund provided the table of various capacitor dielectrics with appropriate constant α that has been obtained experimentally by measurements. Westerlund [E. Pereira et al., 2004] also described behavior of real inductor. For a general current in the inductor the voltage is

$$V(t) = L\frac{\mathrm{d}^{\alpha}I(t)}{\mathrm{d}t^{\alpha}} \equiv L_0 D_t^{\alpha}I(t) \tag{4.2}$$

where L is the inductance of the inductor and constant α is related to the proximity effect.

Based on the above idea, let us consider an extended coupled Duffing circuit containing fractional derivative shown in Figure 4.1. There are two Duffing circuits connected by a linear resistor. Though the voltage difference on it, two Duffing circuit interact with each other. According to notations in the Figure 4.1, the circuit equations can be described as below:

$$\begin{aligned}
D_t^{\alpha}\varphi_1 &= V_1 + e_1(t) \\
D_t^{\alpha}\varphi_2 &= V_2 + e_2(t) \\
C_1 D_t^{\beta}V_1 &= -g_1 V_1 - i_1 + G(V_2 - V_1) \\
C_2 D_t^{\beta}V_2 &= -g_2 V_2 - i_2 + G(V_1 - V_2)
\end{aligned} \tag{4.3}$$

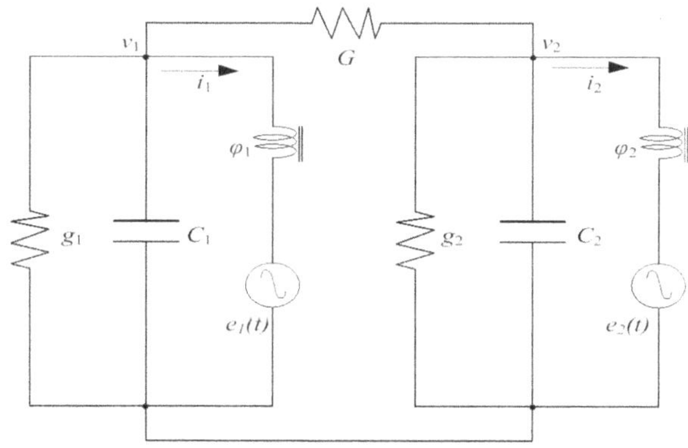

Figure 4.1 A coupled Duffing circuit

where V_1 and V_2 are the voltages cross the capacitors C_1 and C_2, and i_1 and i_2 denote the current through the inductances φ_1 and φ_2, G represents the strength of coupling, D_t^{α} and D_t^{β} are the fractional derivative in the Caputo sense is defined by

Appendix A, i.e.

$$D_t^\mu u(t) = \frac{1}{\Gamma(m-\mu)} \int_0^t (t-\tau)^{m-\alpha-1} \frac{\mathrm{d}^m}{\mathrm{d}\tau^m} u(\tau)\mathrm{d}\tau$$

for $m-1 < \mu \le m, m \in N, t > 0, u \in C_{-1}^m$.

In this circuit, we assume these two capacitors are same and units (i.e. $C_1 = C_2 \overset{\Delta}{=} 1$), the characteristic of the cubic nonlinear inductors can be expressed mathematically

$$i_k = h_1\varphi_k + h_3\varphi_k^3, k = 1,2. \tag{4.4}$$

The external forces of coupled Duffing circuit are chosen as sinusoidal voltage sources with a phase-lift θ and a direct voltage component q_0, that is, we can describe $e_1(t)$ and $e_2(t)$ as:

$$\begin{aligned} e_1(t) &= f_0 + f\sin(\omega t) \\ e_2(t) &= f_0 + f\sin(\omega t + \theta) \end{aligned} \tag{4.5}$$

Introducing the following dimensionless parameters

$$\varphi_1 = u_1, \varphi_2 = u_2, V_1 = v_1, V_2 = v_2, k_1 = g_1, k_2 = g_2, \delta = G,$$

and substituting the above dimensionless parameters and Eqs. (4.4), (4.5) into Eq. (4.3) yields

$$D_t^\alpha u_1 = v_1 + e_1(t) \tag{4.6a}$$

$$D_t^\alpha u_2 = v_2 + e_2(t), \tag{4.6b}$$

$$D_t^\beta v_1 = -h_1 u_1 - h_3 u_1^3 - k_1 v_1 - \delta(v_1 - v_2), \tag{4.6c}$$

$$D_t^\beta v_2 = -h_1 u_2 - h_3 u_2^3 - k_2 v_2 - \delta(v_2 - v_1), \tag{4.6d}$$

$$e_1(t) = f_0 + f\sin(\omega t) \tag{4.6e}$$

$$e_2(t) = f_0 + f\sin(\omega t + \theta) \tag{4.6f}$$

In this section, we discuss the steady state fundamental resonance response generating from fractional order coupled Duffing circuit equation (4.6) and illustrate the effects of parameters on the resonance response.

4.1.3 Steady State Response

For analyzing the dynamic behavior of the system whose equation of motion is given by Eqs. (4.6), an approximate solution corresponding to the steady state response in the region of the primary resonance is sought. We firstly assume the harmonic balance solutions of variables $u_1(t)$ and $u_2(t)$ to be

$$\begin{cases} u_1(t) = a_0 + a_1\cos(\omega t) + b_1\sin(\omega t) \\ u_2(t) = c_0 + c_1\cos(\omega t) + d_1\sin(\omega t) \end{cases} \tag{4.7}$$

where $a_0, a_1, b_1, c_0, c_1, d_1$ are the unknown Fourier coefficients and the frequency ω is determined by the external voltage source.

Substituting the Eq. (4.7) into the Eqs. (4.6a, 4.6b, 4.6e, 4.6f), the variables $v_1(t)$ and $v_2(t)$ easily be obtained as follows:

$$\begin{cases} v_1(t) = D_t^\alpha[a_0 + a_1\cos(\omega t) + b_1\sin(\omega t)] - f_0 - f\sin(\omega t) \\ v_2(t) = D_t^\alpha[c_0 + c_1\cos(\omega t) + d_1\sin(\omega t)] - f_0 - f\sin(\omega t + \theta) \end{cases} \tag{4.8}$$

Then, substituting Eqs. (4.7) and (4.8) into Eqs. (4.6c, 4.6d) and equating constant terms and the coefficients of the terms containing $\cos(\omega t)$ and $\sin(\omega t)$ separately to zero results in the following harmonic balance equations:

$$\begin{cases} R_{0,i}(a_0,a_1,b_1,c_0,c_1,d_1) = \dfrac{2}{\pi}\int_0^\pi \Phi_i(D_t^{\alpha+\beta}u_i, D_t^\alpha u_1, D_t^\alpha u_2, u_i, \lambda)\,d\tau = 0, \\[2mm] R_{1,i}^c(a_0,a_1,b_1,c_0,c_1,d_1) = \dfrac{2}{\pi}\int_0^\pi \Phi_i(D_t^{\alpha+\beta}u_i, D_t^\alpha u_1, D_t^\alpha u_2, u_i, \lambda)\cos\tau\,d\tau = 0, \\[2mm] R_{1,i}^s(a_0,a_1,b_1,c_0,c_1,d_1) = \dfrac{2}{\pi}\int_0^\pi \Phi_i(D_t^{\alpha+\beta}u_i, D_t^\alpha u_1, D_t^\alpha u_2, u_i, \lambda)\sin\tau\,d\tau = 0, i = 1,2. \end{cases} \tag{4.9}$$

where

$$\Phi_1 = D_t^{\alpha+\beta}u_1 - D_t^\beta[f_0 + f\sin(\omega t)] + h_1 u_1 + h_3 u_1^3 + k_1[D_t^\alpha u_1 - f_0 - f\sin(\omega t)]$$
$$+\delta[D_t^\alpha(u_1 - u_2) + f\sin(\omega t + \theta) - f\sin(\omega t)],$$

$$\Phi_2 = D_t^{\alpha+\beta}u_2 - D_t^\beta[f_0 + f\sin(\omega t + \theta)] + h_1 u_2 + h_3 u_2^3 + k_2[D_t^\alpha u_2 - f_0 - f\sin(\omega t + \theta)]$$
$$+\delta[D_t^\alpha(u_2 - u_1) - f\sin(\omega t + \theta) + f\sin(\omega t)].$$

Carrying out the integration yields the nonlinear harmonic balance equations

$$[K(q,\lambda)]\{q\} = \{R\} = 0 \tag{4.10}$$

where $\{R\} = [R_{0,1}, R_{1,1}^c, R_{1,1}^s, R_{0,2}, R_{1,2}^c, R_{1,2}^s]^T$ is the vector of residues to be annihilated, $\{q\} = [2a_0, a_1, b_1, 2c_0, c_1, d_1]^T$ is the Fourier coefficient vectors of the response, and

$[K(q,\lambda)]$ is the total stiffness matrix. Using the sum and product formulae for the trigonometric functions, the expression of steady state harmonic balance equations can be obtained.

In this study, we assume that the $k_1 = k_2$, then the system Eqs. (4.6) are invariant under the coordinate transformations

$$[u_1(t), v_1(t)] \mapsto [u_2(t), v_2(t)]$$

or

$$[u_1(t), v_1(t)] \mapsto [u_2(t+\pi), v_2(t+\pi)]$$

When phase-shift $\theta = 0$ or $\theta = \pi$ i.e. the external forces e_1 and e_2 are under the conditions of in-phase sinusoidal voltage source or anti-phase sinusoidal voltage source. Correspond to the steady state responses (4.7) and (4.8), the harmonic balance equation (4.10) is invariant under the transform

$$(a_0, a_1, b_1, c_0, c_1, d_1) \mapsto (c_0, c_1, d_1, a_0, a_1, b_1)$$

or

$$(a_0, a_1, b_1, c_0, c_1, d_1) \mapsto (c_0, -c_1, -d_1, a_0, -a_1, -b_1)$$

This indicates that there always exists a pair of symmetric solutions when the amplitudes $A_1 \neq C_1$. We call this type of solution as an inversely symmetrical solution which the number of group orbits of this type of solution is two, in other words, this type of solution have a symmetric partner.

Especially, if

$$a_0 = c_0, a_1 = c_1, b_1 = d_1$$

or

$$a_0 = c_0, a_1 = -c_1, b_1 = -d_1$$

the number of orbits of above solutions becomes one, we call this type of solution as a completely symmetrical periodic solution. where, $A_1 = \sqrt{a_1^2 + b_1^2}, C_1 = \sqrt{c_1^2 + d_1^2}$.

4.1.4 Effects of the System Parameters on the Resonance Response and Discussions

By means of technique of residue harmonic balance in combination with polynomial homotopy continuation which described in chapter 1, all the isolated solutions and the corresponding improved corrections of the harmonic balance equations (4.10) can be obtained. In this section, the results obtained are illustrated and compared with the results of direct numerical integration of the Eq. (4.6) of motion. The influence of commensurate fractional order is examined by the frequency versus response and excitation amplitude versus response curves under different commensurate fractional orders. The steady state responses under different incommensurate fractional order systems are also illustrated and compared by curves of fractional order versus amplitude. For examining the influence of coupling intensity parameter on equation of motion, we illustrate the steady state frequency amplitude curves for three different coupled parameters under anti-phase sinusoidal source in a final subsection. In Figures 4.2–4.7, solid and broken lines denote stable and unstable branches respectively, dots are results from numerical integration.

4.1.4.1 Frequency versus response amplitudes under different commensurate fractional orders

In this subsection, the relations of frequency versus response amplitudes under different fractional orders are illustrated and compared with the results of direct numerical integration of the equation of motion. It should be emphasized that the system parameters are chosen in such way that the response is predominantly at the same frequency as the harmonic excitation, and that all other Fourier components can be neglected in comparison to the bias term and the fundamental harmonic term. In Figure 4.2, the system parameters as $h_1=1$, $h_3=1$, $k_1=0.2$, $k_2=0.2$, $f_0=1$,

$f=0.5$, $\delta=0.1$, $\theta=0$ are fixed and only the fractional order is different. Owing to the inversely symmetrical periodic solutions are unstable, we only show the completely symmetrical periodic solutions. Figure 4.2 illustrates the frequency versus fundamental amplitude curves for fractional order $\alpha=0.9$, $\beta=0.9$, $\alpha=0.99$, $\beta=0.99$ and $\alpha=1$, $\beta=1$ respectively. For the case of $\alpha=0.9$, $\beta=0.9$, we can see that the amplitude increases at first, then decreases with the increasing of frequency and only one stable solution exist. When frequency increases to 1.29, the amplitude arrives at maximum peak amplitude. For the cases of $\alpha=0.99$, $\beta=0.99$ and $\alpha=1$, $\beta=1$, we can see (i) when excitation frequency increases from 0 to 1.560575 and 1.562465 for fractional orders $\alpha=0.99$, $\beta=0.99$ and $\alpha=1$, $\beta=1$ respectively, only one stable harmonic solution exists and the fundamental amplitude increases monotonously. (ii) The amplitudes appear hysteresis phenomena which take place at excitation frequency ω is in between 1.560575 and 1.967025, 1.562465 and 2.39447 corresponding to the fractional order $\alpha=0.99$, $\beta=0.99$ and $\alpha=1$, $\beta=1$ respectively. Although the response curves are roughly of similar shape in Figure 4.2, the curvatures at the turning points are different for the above cases. In Figure 4.3, we illustrate the frequency versus response curves of bias terms A_0 or C_0 for fractional order $\alpha=0.9$, $\beta=0.9$; $\alpha=0.99$, $\beta=0.99$ and $\alpha=1$, $\beta=1$, respectively. From Figures 4.2–4.3, it is observed that there is the considerable difference for different fractional order systems. The obtained harmonic solutions are in excellent agreement with numerical results.

4.1.4.2 Excitation amplitude versus response amplitudes under different commensurate fractional orders

In this subsection, we focus on the effects of different commensurate fractional orders to steady state response induced by the change in the amplitude of sinusoidal driving. The system parameters $h_1=0$, $h_3=1$, $k_1=0.2$, $\omega=1$, $k_2=0.2$, $f_0=0$, $\delta=0.1$, $\theta=\pi$ are taken and only the fractional order is different. The excitation

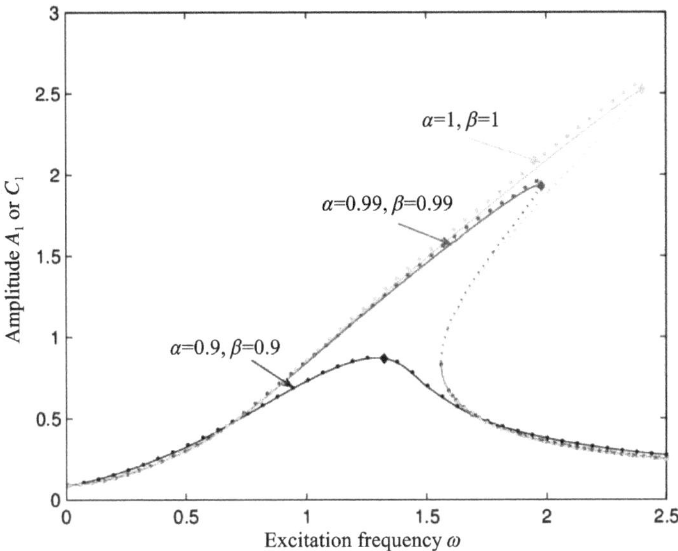

Figure 4.2 Frequency versus response amplitude of harmonic terms A_1 for fractional order $\alpha = 0.9$, $\beta = 0.9$; $\alpha = 0.99$, $\beta = 0.99$ and $\alpha = 1$, $\beta = 1$, " ◆ " peak amplitude

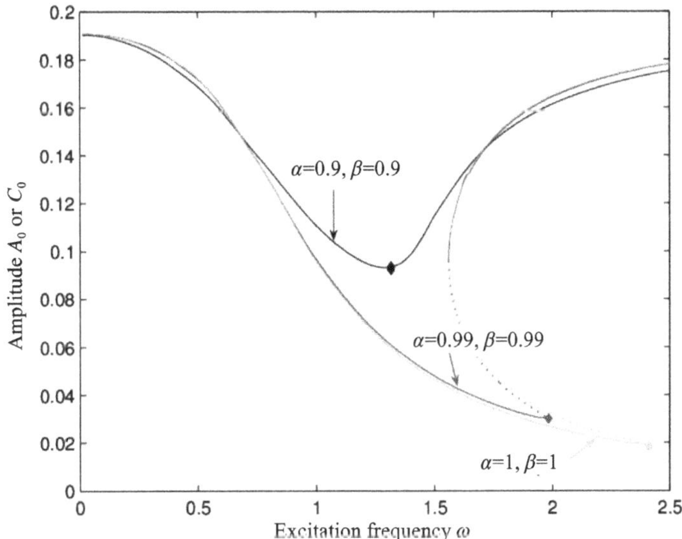

Figure 4.3 Frequency versus response amplitude of bias terms A_0 or C_0 for fractional order $\alpha = 0.9$, $\beta = 0.9$; $\alpha = 0.99$, $\beta = 0.99$ and $\alpha = 1$, $\beta = 1$," ◆ " peak amplitude

amplitude versus response amplitudes are illustrated in Figures 4.4 and 4.5 with commensurate fractional order $\alpha = 0.9$, $\beta = 0.9$ and $\alpha = 0.98$, $\beta = 0.98$, respectively. Figures 4.6 and 4.7 show the comparisons between the present stable analytical solutions and the numerical integration results for $q = 0.45$ corresponding to Figures

4.4 and 4.5 respectively. From Figures 4.4 and 4.5, it can be observed that there is a considerable difference on the excitation amplitude versus steady state amplitude curves for two different commensurate fractional order cases. For the case of the fractional order $\alpha=0.9$, $\beta=0.9$, there is only one completely symmetrical periodic solution when the excitation amplitude is smaller. The completely symmetrical solution loses stability and bifurcates a new inversely symmetrical periodic solution as the excitation amplitude increases to point $f_1(f\approx0.58398)$. In point $f_2(f\approx0.62232)$, the amplitudes of the inversely symmetrical solution disappear and fold, and the completely symmetrical solution becomes stability again. For the case of the fractional order $\alpha=0.98$, $\beta=0.98$, the steady state response becomes more complicated than mentioned above and multiple-valued stable solutions, indicating the occurrence of hysteresis and jump phenomena, are observed analytically. From Figure 4.5, it can be observed that (i) there is only one stable completely symmetrical solution when the excitation amplitude is less than the value in point $f_1(f\approx0.385762)$. (ii) At point f_1, two fold bifurcations occur and bifurcate two pairs of inversely symmetrical periodic solutions, in which two are stable and the other two are unstable. That is, the system exists five different periodic solutions when the excitation amplitude increasing from point f_1 to turning point $f_2(f\approx0.4714)$ and three stable solutions coexist where one is completely symmetrical, the other two are inversely symmetrical. (iii) Two pitchfork bifurcations of limit cycle occur at points $f_3(f\approx0.4806)$ and $f_4(f\approx0.496322)$. There are three completely symmetrical periodic solutions when the excitation amplitude is between the turning points f_2 and $f_5(f\approx0.503647)$. (iv) As the excitation amplitude further increases, two pairs of inversion symmetrical periodic solutions mentioned above fold at point $f_6(f\approx0.547325)$. That is, two new fold bifurcations occur at point f_6. What is more, there is only one completely symmetrical periodic solution again when excitation amplitude increases form point f_6 to 0.65. To further verify these predictions analytically, Figure 4.6 shows a comparisons of the phase portrait about the variable $u_1(t)$ at point $f=0.45$ for fractional order $\alpha=0.9$, $\beta=0.9$. The corresponding analytical expression is as below.

$$u_1(t) = -0.49787\cos(t) + 0.14691\sin(t) - 0.0011124\cos(3t) + 0.0047916\sin(3t)$$

Figure 4.7 shows the phase portrait for coexistence of three stable solutions $u_1(t)$ at point $f=0.45$ for fractional order $\alpha=0.98$, $\beta=0.98$ and all analytical solutions are tabulated in Table 4.1. The analytical solutions are in excellent agreement with numerical integration results and it should be provided that when two or more stable solutions coexist, the initial conditions determine which steady state solution is picked up.

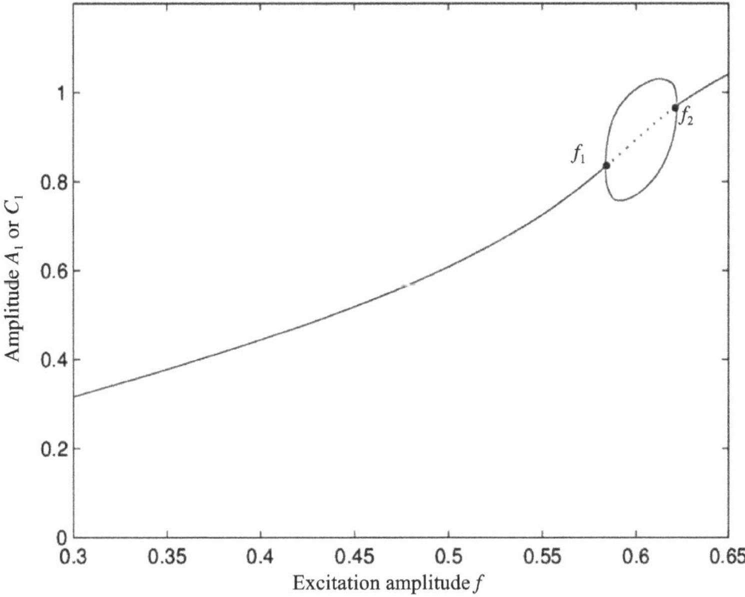

Figure 4. 4 Excitation amplitude versus response amplitudes of harmonic terms A_1 or C_1 for fractional order $\alpha=0.9$, $\beta=0.9$

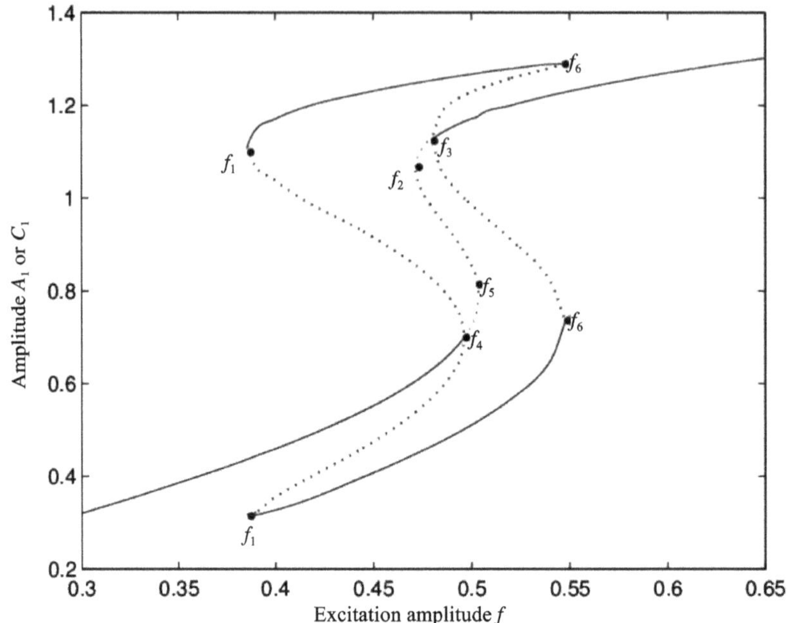

Figure 4. 5 Excitation amplitude versus response amplitudes of harmonic terms
A_1 or C_1 for fractional order $\alpha = 0.98$, $\beta = 0.98$

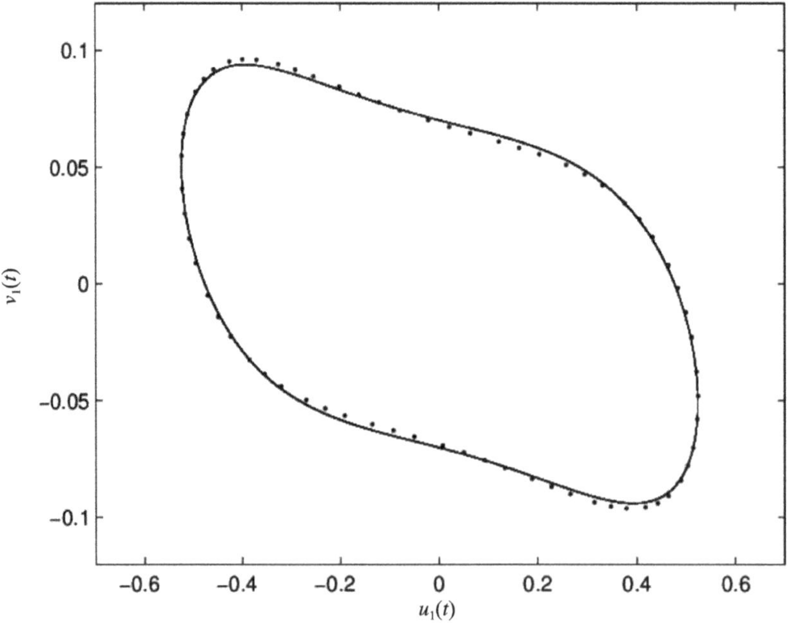

Figure 4. 6 Phase portraits of stable solution corresponding to
Figure 4.4 with $f = 0.45$

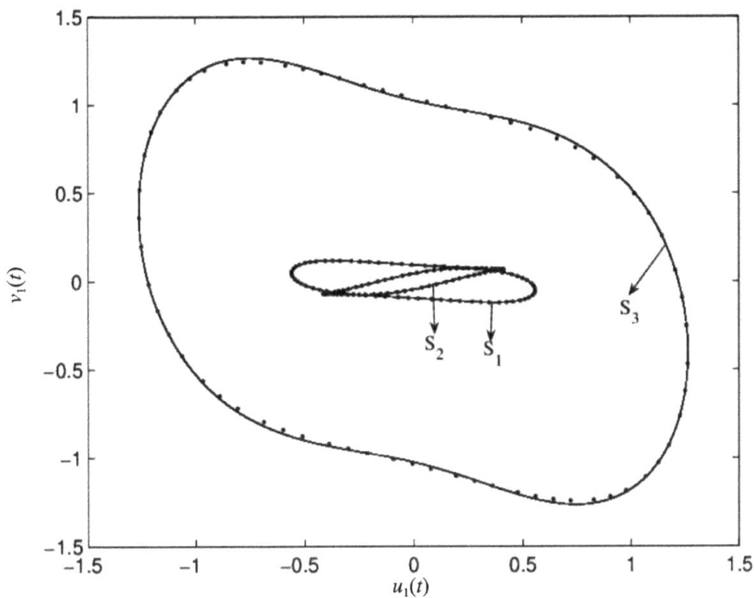

Figure 4.7 Phase portraits of stable solutions corresponding to Figure 4.5 with
f = 0.45 and three stable solutions coexist

Table 4.1 Analytical solutions for fractional order $\alpha = 0.98$, $\beta = 0.98$ and $f = 0.45$

	Stable (S_1)	Stable (S_2)	Stable (S_3)	Unstable (S_4)	Unstable (S_5)
$\cos \omega t$	-0.54841	-0.40707	0.06079	-0.80308	-0.47480
$\sin \omega t$	0.07941	-0.05640	1.19672	0.38929	0.07324
$\cos 3\omega t$	-0.00412	-0.00401	-0.02413	-0.00058	-0.00185
$\sin 3\omega t$	0.00307	0.00063	-0.06181	0.02389	0.00186

4.1.4.3 Incommensurate fractional order system and effect

In this subsection, we consider the case of incommensurate fractional order system and illustrate the difference of different fractional orders on steady state responses. The system parameters $h_1 = 1$, $h_3 = 1$, $k_1 = 0.2$, $k_2 = 0.2$, $m = 2$, $f_0 = 0$, $f = 0.5$, $\delta = 0.1$, $\theta = 0$ are fixed. Figures 4.8a and 4.8b show the relation of fractional order versus amplitude of steady state induced by the change in the fractional order for three cases, $Case_1$: Fix $\beta = 1$ and vary $\alpha \in [0.9,1]$; $Case_2$: Fix $\alpha = 1$ and vary $\beta \in [0.9,1]$; Case3: vary α, $\beta \in [0.9,1]$ simultaneously. From Figure 4.8a, it can be observed that there exist one completely symmetrical solution with smaller fundamental amplitude for each case and these three fundamental amplitudes decrease as the fractional order increases, but their slopes are different. When the fractional order increases to some values, two completely symmetrical and a pair of inversely symmetrical periodic solutions appears one after another. In Figure 4.8b, solid lines, dashed lines and dashed-dot lines denote the stable completely symmetrical, unstable completely symmetrical and unstable inversely symmetrical periodic solutions. For $Case_1$, when fractional order α increases to $\alpha_1 (\alpha \approx 0.981425)$, two completely symmetrical periodic solutions occur, the upper branch is stable and other branch is unstable. Further, as the fractional order increases to $\alpha_2 (\alpha \approx 0.98361)$, a pair of inversely symmetrical periodic solution occurs and two branches are unstable. Similar to $Case_1$, the other key bifurcation points are $\beta_1 (\beta \approx 0.9832665)$, $\beta_2 (\beta \approx 0.984915)$, $\alpha_3 (\beta_3)(\alpha = \beta = 0.9912)$, $\alpha_4 (\beta_4)$ $(\alpha = \beta = 0.992145)$ for $Case_2$ and $Case_3$ respectively.

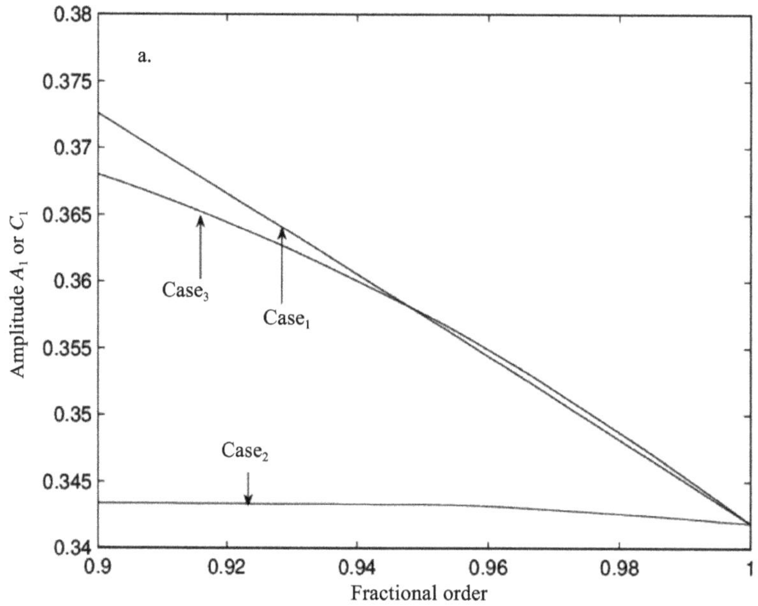

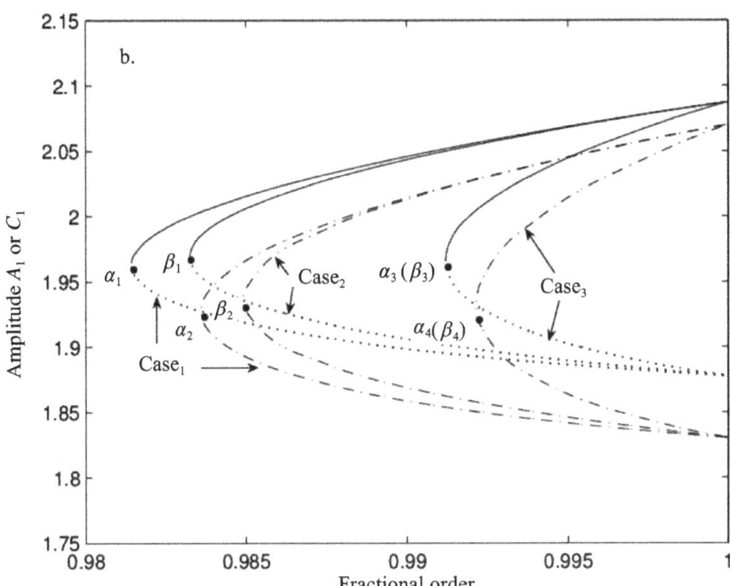

Figure 4. 8 Fractional order versus amplitudes of harmonic terms A_1 or C_1 for three cases, Case$_1$: Fix $\beta=1$ and vary $\alpha\in[0.9,1]$; Case$_2$: Fix $\alpha=1$ and vary $\beta\in[0.9,1]$; Case$_3$: vary simultaneously α, $\beta\in[0.9,1]$

4.1.4.4 Effect of coupling intensity parameter under anti-phase sinusoidal source

We assume that the external forces are anti-phase sinusoidal voltage source, and steady state responses are illustrated to examine the influence of coupled parameteron the different shapes curves of the frequency versus amplitude. In Figures 4.9–4.11, we fix the system parameters as $h_1=1$, $h_3=1$, $k_1=0.2$, $k_2=0.2$, $f_0=1$, $f=0.5$, $\theta=\pi$, $\alpha=0.95$, $\beta=0.95$. The solid line, dashed lines and dots denote the coupling intensity parameter $\delta=0$, $\delta=0.5$ and $\delta=1$ respectively. When consider the case of $\delta=0$ in Eqs. (4.6), that are the connection of two parts is broken, the system is turning to two isolated Duffing circuits. From Figures 4.9 and 4.10, there is a considerable difference among three frequencies versus amplitudes of bias term and fundamental solutions. Figure 4.11 shows a comparisons of the phase portrait about the variable $u_1(t)$ at point $\omega=1.5$. The corresponding analytical expressions are tabulated in Table 4.2. It is clear that the present analytical solutions are in excellent agreement with numerical integration results.

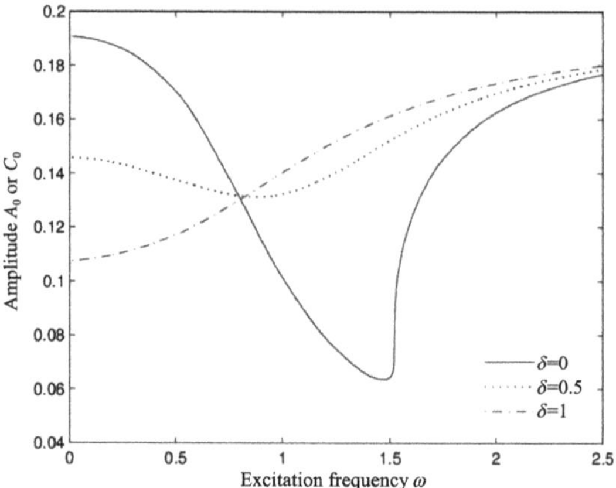

Figure 4.9 Frequency versus response amplitudes of bias terms A_0 for fractional order
$\alpha = 0.95$, $\beta = 0.95$ and for different values of the coupling intensity parameter

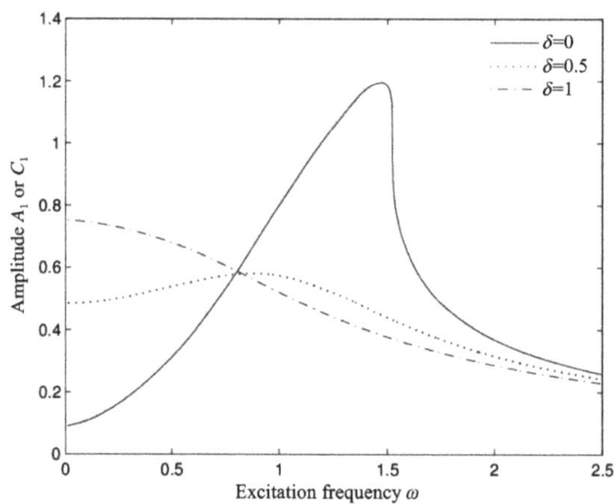

Figure 4.10 Frequency versus response amplitudes of harmonic terms A_1 or C_1 for fractional order
$\alpha = 0.95$, $\beta = 0.95$ and for different values of the coupling intensity parameter

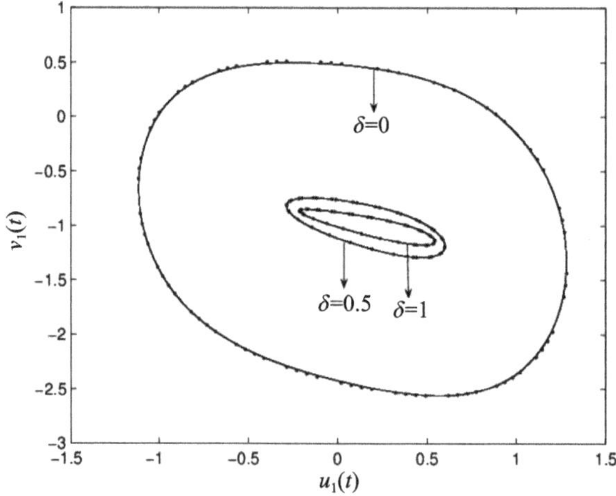

Figure 4.11 Phase portraits of stable solutions corresponding to Figures 4.9 and 4.10 with
$\omega = 1.5$ solid lines: analytical solution, dotted lines: numerical result

Table 3.2 Analytical solutions for fractional order $\alpha = 0.95$, $\beta = 0.95$ and $\omega = 1.5$

	$\delta = 0$		$\delta = 0.5$		$\delta = 1$	
	$u_1(t)$	$u_2(t)$	$u_1(t)$	$u_2(t)$	$u_1(t)$	$u_2(t)$
Constant	0.05783	0.05783	0.15157	0.15157	0.16093	0.16093
$\cos \omega t$	-0.32924	0.32924	-0.39006	0.39006	-0.35165	0.35165
$\sin \omega t$	1.12217	-1.12217	0.20480	-0.20480	0.13723	-0.13723
$\cos 2\omega t$	-0.02377	-0.02377	0.00209	0.00209	0.00265	0.00265
$\sin 2\omega t$	-0.00468	-0.00468	-0.00615	-0.00615	-0.00426	-0.00426
$\cos 3\omega t$	0.01586	-0.01586	0.00046	-0.00046	0.00017	-0.00017
$\sin 3\omega t$	-0.02305	0.02305	0.00122	-0.00122	0.00072	-0.00072

4.1.5 Concluding Remarks

Classical electrical circuits, consist of resistors and capacitors, are described by integer-order models. However, circuits may have the so-called fractance that represents an electrical element with fractional order impedance. Therefore, the fractional calculus existing in the circuit is a factor, for one, to consider dynamical behaviors. This section extends a classical coupled Duffing circuit. The new fractional order Duffing system consisting of two identical periodic forced circuits coupled by a linear resistor is presented. The fundamental resonance responses of system under different commensurate and incommensurate fractional orders have been investigated in detail using the residue harmonic balance in combination with polynomial homotopy continuation technique.

To the best of our knowledge, the present study may be the first one by introducing the fractional derivative to coupled Duffing circuit and illustrate the effects of different fractional orders on the various kinds of response curves. The present section not only provides the analytical form of fundamental harmonic response with a high accuracy, but also examines various bifurcations in term of the parameters. In more specific terms, it was found that:

(i) The frequency amplitude curves not only have different shapes but also exist significant difference for fundamental amplitude value under different commensurate fractional order, which are shown in Figures 4.2 and 4.3. Besides, the fundamental peak amplitude values gradually increases with the increasing of the commensurate fractional order.

(ii) The different shapes of excitation amplitude versus response amplitudes of fundamental term are found for different commensurate fractional orders those are shown in Figures 4.4-4.7. For instance, multiple-valued solutions, indicating the occurrence of jump phenomena and bifurcation are observed analytically and confirmed numerically for $\alpha = 0.98$, $\beta = 0.9$.

(iii)The different incommensurate fractional orders affect the bifurcation points of frequency versus fundamental amplitude curves that are illustrated in Figure 4.8.

(iv)The different coupled parameter has bigger influence on the frequency amplitude curves under excitation of anti-phase sinusoidal voltage source. Specially, when $\delta = 0$, the shape and peak amplitude are completely different from other two type cases $\delta = 0.5$ and $\delta = 1$.

(v) The approximate solutions corresponding to steady state resonance responses are in excellent agreement with numerical integration results. The above results have potential applications on many general coupled fractional order systems.

4.2 Hard–spring Bistability and Effect of System Parameters in a Two–degree–of–freedom Vibration System with Damping Modelled by a Fractional Derivative

4.2.1 Introduction

The jump and bifurcation phenomena for geometrical nonlinear mechanical system subject to high amplitude vibration are often dangerous and undesirable. The dynamic absorber is sometimes required to reduce the vibration of system because it has the advantages of low cost and simple operation at one modal frequency [Soom & Lee, 1983; Jordanov & Cheshankov, 1988; Song et al., 2003]. In the domain of many mechanical vibration systems, the coupled nonlinear vibration of such systems can be reduced to second order nonlinear differential equations, but so far there are no efficient methods available for a nonlinear structural frequency response analysis on a large scale. For weakly or mildly nonlinear vibration systems, many effective analytical methods have been suggested, such as the averaging method, multiple time scale method, and perturbation method, etc. For instance, the method of averaging was applied to investigate the steady state oscillation and stability of nonlinear dynamic vibration absorbers. The study found that proper selection of the system parameter would result in insubstantial improvements of nonlinear absorbers and avoid dangerous effects that are likely to occur due to the presence of the nonlinearities [Natsiavas, 1992]. EI-Bassiouny [2005] used the multiple time scale to study the effects of quadratic and cubic nonlinearities in elastomeric material dampers on torsional vibration control. He showed that the elastomeric damper reduced the vibration of the crankshaft effectively. Sayed et al. [2011] used the multiple time scales perturbation method to investigate vibration of a two degree-of-freedom vibration system including quadratic and cubic nonlinearities subjected to external and parametric excitation forces. The vibration of the main system can be controlled applying nonlinear absorber. They extracted all possible resonance cases and studied numerically the effects of different parameters of the system.

Recently, the fractional calculus [Podlubny, 1999; Debnath, 2003; Kilbas et al., 2006; Rossikhin & Shitikova, 2010; Caponetto et al., 2010; Baleanu et al., 2012; Uchaikin, 2013] has attracted fairly broad research activity, including the viscoelasticity [Rossikhin & Shitikova, 2010], quantum mechanics [Laskin, 2000], mechanics of non-Hamiltonian systems [Tarasov, 2006],physical kinetics [Saichev

& Zaslavsky, 1997] and so on [Richard et al., 2011]. Due to the occurrence of fractional oscillators in dynamical applications, various kinds of vibration systems with damping modelled by a fractional derivative have attracted more and more attention, see for instance [Petras, 2011] and references therein. Many phenomena are found in the integer order dynamic system, such as different kinds of resonance, hysteresis, bifurcation and then chaos can also be exhibited in the fractional derivative system [Debnath, 2003; Sheu et al., 2007; Rossikhin & Shitikova 2010; Leung et al., 2012, 2013]. As a result, by introducing fractional derivative damping into mechanical system to study the effects of fractional derivative order, linear and nonlinear dampings on the system response is one of the original contributions to the fractional system subject to a nonlinear absorber.

The harmonic balance method [Urabe, 1965; Luo, 2014] is an attractive solution method for computing periodic responses. It does not involve the assumption of small parameters as is customarily required by the perturbation and averaging methods. In the method, the periodic solution of the response is expanded into the truncated Fourier series with a finite number of harmonic terms and substituted into the governing equations. The corresponding harmonic coefficients are then equated to zero to give a set of algebraic equations. The classical harmonic balance is simple in its principle, but it may be cumbersome or even impracticable, as stated by [Peng et al., 2008] when the system contains complex nonlinearities and when a large number of harmonics is required. To improve the harmonic balance method, the polynomial homotopy continuation technique [Li, 1997; Sommese & Wampler, 2005; Wu, 2005, 2006] is introduced to solve the polynomial system of equations generated from the harmonic balance method for nonlinear vibration system, so that bifurcation analysis can be carried out in this study.

This work aims to introduce polynomial homotopy continuation to study the vibration problems of steady state response and bifurcation of a two-degree-of freedom fractional vibration system including quadratic and cubic nonlinearities subjected to external and parametric excitation forces by updating the widely applicable model within [Sayed et al., 2011]. After this introduction, we begin in section 4.2.2 by presenting a two-degree-of freedom fractional vibration system including quadratic and cubic nonlinearities subjected to external and parametric excitation forces. In section 4.2.3, the harmonic balance in combination with polynomial homotopy continuation is described to obtain the periodic response of the nonlinear fractional system. Using this method, the steady state response and the hard-spring bistability scenario are analyzed in section 4.2.4. In section 4.2.5, the effects of fractional derivative order, linear and nonlinear damping coefficients on the system responses are illustrated and discussed. Finally, the conclusion of this study is outlined in section 4.2.6.

4.2.2 Fractional vibration system and fractional derivative

Using a nonlinear tuned mass absorber connected to the main system, a model of a two degree-of-freedom oscillator under consideration is given in [Sayed et al., 2011], from the principles of the mechanics, the derived equations of motion are written in the forms

$$\ddot{x}_1 + \omega_1^2 x_1 + k_1 x_1^2 + k_2 x_1^3 + \xi_1 \dot{x}_1 + \xi_2 \dot{x}_1 x_1^2 + \xi_3(\dot{x}_1 - \dot{x}_2) + k_3 x_2$$
$$+k_4(x_1 - x_2)^2 + k_5(x_1 - x_2)^3 = f_1 \cos(\omega t) + f_2 x_1 \cos(\omega t) \qquad (4.11)$$
$$\ddot{x}_2 + \omega_2^2(x_2 - x_1) + k_6(x_2 - x_1)^2 + k_7(x_2 - x_1)^3 + \xi_4(\dot{x}_2 - \dot{x}_1) = 0$$

The detail and schematic diagram of the model with absorber can be found in [Sayed et al., 2011].

Recently, due to the fractional derivative can provide an excellent instrument for the description of memory and hereditary properties of various materials and process, many researches [Rossikhin & Shitikova, 2010; Sheu et al., 2007; Leung & Guo, 2013, 2014] have been shown that fractional differential equations can better model many real world physical systems and engineering problems. In this study, we assume that the above model vibrates in a viscous medium whose viscous features are defined by fractional derivative which describes the processes of internal friction proceeding in the mechanical system. In this study, we introduce a fractional order time derivative in Eq. (4.11) to consider a slightly different version of the fractional vibration system, and the standard form of a two-degree-freedom fractional vibration system under consideration is written as

$$\ddot{x}_1 + \omega_1^2 x_1 + k_1 x_1^2 + k_2 x_1^3 + \xi_1 D_t^\lambda x_1 + \xi_2 D_t^\lambda x_1 x_1^2 + \xi_3 (D_t^\lambda x_1 - D_t^\lambda x_2) + k_3 x_2$$
$$+k_4 (x_1 - x_2)^2 + k_5 (x_1 - x_2)^3 = f_1 \cos(\omega t) + f_2 x_1 \cos(\omega t) \tag{4.12}$$
$$\ddot{x}_2 + \omega_2^2 (x_2 - x_1) + k_6 (x_2 - x_1)^2 + k_7 (x_2 - x_1)^3 + \xi_4 (D_t^\lambda x_2 - D_t^\lambda x_1) = 0$$

Where x_1 and x_2 denote normalized displacement function of main system and absorber, respectively. ξ_1 and ξ_2 are the linear and nonlinear damping coefficients of the main system, respectively. ξ_3 and ξ_4 are the coupling damping coefficients, ω is forcing frequency, ω_1, ω_2, are natural frequencies, k_3 is linear stiffness coefficient, f_1 is external excitation amplitude, f_2 is parametric excitation amplitude, $k_i (i = 1, 2, \cdots, 7)$ are nonlinear stiffness coefficients. D_t^λ is the fractional derivative of order $\lambda (0 < \lambda \le 1)$ and in the sense of Caputo definition [Podlubny, 1999] is defined as

$$D_t^\lambda x(t) = \frac{1}{\Gamma(m - \lambda)} \int_0^t (t - \tau)^{m - \lambda - 1} \frac{d^m}{d\tau^m} x(\tau) d\tau \tag{4.13}$$

$$m - 1 < \lambda \le m, m \in N, t > 0, x(\tau) \in C_{-1}^m.$$

In general, three definitions are often used for the general fractional derivative. Besides of the above Caputo definition, the other two definitions, Grunwald-Letnikov derivative and Riemann-Liouville derivative, are as follow respectively

$$D_t^\lambda x(t) = \frac{1}{\Gamma(m - \lambda)} \int_0^t (t - \tau)^{m - \lambda - 1} \frac{d^m}{d\tau^m} x(\tau) d\tau \tag{4.14}$$

and

$$_{0}^{RL}D_t^\lambda x(t) = \frac{1}{\Gamma(m - \lambda)} \frac{d^m}{dt^m} \left(\int_0^t (t - \tau)^{m - \lambda - 1} x(\tau) d\tau \right) \tag{4.15}$$

$$m - 1 < \lambda \le m$$

It has been known that if theis (m-1)-times continuously differentiable and $x'''(t)$ is integrable, then

$$^{GL}_0D^\lambda_t x(t) = {}^{RL}_0D^\lambda_t x(t) = D^\lambda_t x(t) + \sum_{k=0}^{m-1}\frac{t^{k-\lambda}}{\Gamma(k-\lambda+1)}x^{(k)}(0^+), \quad m-1<\lambda\le m \tag{4.16}$$

Thus, whentends to infinity we have

$$^{GL}_0D^\lambda_t x(t) = {}^{RL}_0D^\lambda_t x(t) = D^\lambda_t x(t) \tag{4.17}$$

That is, these three aforementioned definitions of fractional-order derivative are equivalent in the sense of steady state response. For simplicity, we will use Grunwald-Letnikov derivative format instead of Caputo definition in the process of numerical simulation in this study.

4.2.3 Steady state response and solution procedure

4.2.3.1 Steady state response and harmonic balance analysis

For analyzing the dynamic behavior of the system whose equation of motion is given by Eq. (4.12), an approximate solution corresponding to the steady state response in the region of the primary resonance is sought. We consider the analytical solution of period motion as

$$x_1(t) = a_0 + \sum_{i=1}^{n}[a_i\cos(i\omega t)+b_i\sin(i\omega t)]$$

$$x_2(t) = c_0 + \sum_{i=1}^{n}[c_i\cos(i\omega t)+d_i\sin(i\omega t)] \tag{4.18}$$

$D^\lambda_t x_1(t)$ and $D^\lambda_t x_2(t)$ can be computed in the sense of steady state response by using contour integration as below:

$$D^\lambda_t x_1(t) = \sum_{i=1}^{n}g_i[(a_i m_1+b_i n_1)\cos(i\omega t)+(b_i m_1-a_i n_1)\sin(i\omega t)]$$

$$D^\lambda_t x_2(t) = \sum_{i=1}^{n}g_i[(c_i m_1+d_i n_1)\cos(i\omega t)+(d_i m_1-c_i n_1)\sin(i\omega t)] \tag{4.19}$$

where $g_i = (i\omega)^\lambda, m_1 = \cos(\dfrac{\lambda\pi}{2}), n_1 = \sin(\dfrac{\lambda\pi}{2})$ and computing detail can be found in Ref. [Guo et al., 2014]. a_0, c_0 are the bias term of main system and absorber respectively, $a_i, b_i, c_i, d_i (i=1,2,\cdots)$ are the i-th harmonic components of main system and absorber respectively. For convenience, we introduce a new time scale $\tau = \omega t$, then Eq. (4.12) and Eq. (4.14) become

$$\omega^2 x_1'' + \omega_1^2 x_1 + k_1 x_1^2 + k_2 x_1^3 + \xi_1\omega^\lambda(D_\tau^\lambda x_1) + \xi_2\omega^\lambda(D_\tau^\lambda x_1)x_1^2 + \xi_3\omega^\lambda(D_\tau^\lambda x_1 - D_\tau^\lambda x_2)$$

$$+ k_3 x_2 + k_4(x_1 - x_2)^2 + k_5(x_1 - x_2)^3 = f_1\cos(\tau) + f_2 x_1\cos(\tau) \tag{4.20}$$

$$\omega^2 x_2'' + \omega_2^2(x_2 - x_1) + k_6(x_2 - x_1)^2 + k_7(x_2 - x_1)^3 + \xi_4\omega^\lambda(D_\tau^\lambda x_2 - D_\tau^\lambda x_1) = 0$$

$$x_1(\tau) = a_0 + \sum_{i=1}^{n}[a_i\cos(i\tau) + b_i\sin(i\tau)]$$

$$x_2(\tau) = c_0 + \sum_{i=1}^{n}[c_i\cos(i\tau) + d_i\sin(i\tau)] \tag{4.21}$$

Substituting Eq. (4.21) into Eq. (4.20) and using the Galerkin procedure and Fourier expression of fractional order derivative, result in the following harmonic balance equations:

$$\left\{\begin{array}{l}
R_{0,i}(a_0,a_1,b_1,\cdots,c_0,c_1,d_1,\cdots) = \dfrac{2}{\pi}\int_0^\pi \Pi_i(x_1'',x_2'',D_\tau^\lambda x_1,D_\tau^\lambda x_2,x_1,x_2,\gamma)\mathrm{d}\tau = 0, \\[2mm]
R_{1,i}^c(a_0,a_1,b_1,\cdots,c_0,c_1,d_1,\cdots) = \dfrac{2}{\pi}\int_0^\pi \Pi_i(x_1'',x_2'',D_\tau^\lambda x_1,D_\tau^\lambda x_2,x_1,x_2,\gamma)\cos(\tau)\mathrm{d}\tau = 0, \\[2mm]
R_{1,i}^s(a_0,a_1,b_1,\cdots,c_0,c_1,d_1,\cdots) = \dfrac{2}{\pi}\int_0^\pi \Pi_i(x_1'',x_2'',D_\tau^\lambda x_1,D_\tau^\lambda x_2,x_1,x_2,\gamma)\sin(\tau)\mathrm{d}\tau = 0, i=1,2,\cdots \\[2mm]
\qquad\qquad\qquad\cdots \\[2mm]
R_{n,i}^c(a_0,a_1,b_1,\cdots,c_0,c_1,d_1,\cdots) = \dfrac{2}{\pi}\int_0^\pi \Pi_i(x_1'',x_2'',D_\tau^\lambda x_1,D_\tau^\lambda x_2,x_1,x_2,\gamma)\cos(\tau)\mathrm{d}\tau = 0, \\[2mm]
R_{n,i}^s(a_0,a_1,b_1,\cdots,c_0,c_1,d_1,\cdots) = \dfrac{2}{\pi}\int_0^\pi \Pi_i(x_1'',x_2'',D_\tau^\lambda x_1,D_\tau^\lambda x_2,x_1,x_2,\gamma)\sin(\tau)\mathrm{d}\tau = 0,
\end{array}\right. \tag{4.22}$$

where

$$\Pi_1 = \omega^2 x'' + \omega_1^2 x_1 + k_1 x_1^2 + k_2 x_1^3 + \xi_1\omega^\lambda(D_\tau^\lambda x_1) + \xi_2\omega^\lambda(D_\tau^\lambda x_1)x_1^2 + \xi_3\omega^\lambda(D_\tau^\lambda x_1 - D_\tau^\lambda x_2) + k_3 x_2$$

$$+ k_4(x_1 - x_2)^2 + k_5(x_1 - x_2)^3 - f_1\cos(\tau) - f_2 x_1\cos(\tau),$$

$$\Pi_2 = \omega^2 x_2'' + \omega_2^2(x_2 - x_1) + k_6(x_2 - x_1)^2 + k_7(x_2 - x_1)^3 + \xi_4\omega^\lambda(D_\tau^\lambda x_2 - D_\tau^\lambda x_1),$$

$$\gamma \stackrel{\Delta}{=} \gamma(\omega,\omega_1,\omega_2,\xi_1,\cdots,\xi_4,k_1,\cdots,k_7,f_1,f_2,\lambda).$$

Carrying out the integration yields the nonlinear harmonic balance equations

$$R\{q,\gamma\} \equiv [K(q,\gamma)]\{q\} + \{F\} = 0 \tag{4.23}$$

where $\{R\} = [R_{0,1}, R_{1,1}^c, R_{1,1}^s, \cdots R_{n,1}^c, R_{n,1}^s, R_{0,2}, R_{1,2}^c, R_{1,2}^s, \cdots, R_{n,2}^c, R_{n,2}^s]^T$ is the vector of residues to be annihilated, $\{q\} = [a_0, a_1, b_1, \cdots, a_n, b_n, c_0, c_1, d_1, \cdots, c_n, d_n]^T$ and $\{F\}$ are the Fourier coefficient vectors of the response and excitation respectively, and $[K(q,\gamma)]$ is the total stiffness matrix. Using the sum and produce formulate for the trigonometric functions, the expression of steady state harmonic balance equations can be obtained. When $n = 1$ in Eq. (4.18), the corresponding harmonic balance equations can be found in APPENDIX C.

4.2.3.2 Basic idea of polynomial homotopy continuation technique

This section performs a brief review of the theory of polynomial homotopy continuation method for solving Eq. (4.23). For convenience, we rewrite nonlinear Eq. (4.23) into the following form

$$\Phi(q) = [\Phi_1(q), \Phi_2(q), \cdots, \Phi_{4n+2}(q)]^T = 0 \tag{4.24}$$

To solve Eq. (4.24), we choose a new simple start system called the auxiliary homotopy function

$$G(q) = [G_1(q), G_2(q) \cdots, G_{4n+2}(q)]^T = 0 \tag{4.25}$$

The auxiliary homotopy function $G(q)$ is known or controllable and is easy to solve. Then we define the homotopy continuation function as

$$H(q,\eta) = (1-\eta)\sigma G(q) + \eta\Phi(q) \tag{4.26}$$

Where η is an arbitrary parameter that varies from 0 to 1. σ is a set of given coefficient. Therefore, we have the following two boundary conditions:

$$H(q,0) = G(q), \quad H(q,1) = \Phi(q). \tag{4.27}$$

Then our aim is to solve the $H(q,\eta) = 0$ instead of $\Phi(q) = 0$ by changing the homotopy parameter η from 0 to 1. By appropriate adjusting the auxiliary homotopy function [Wu, 2005; 2006], the solution of Eq. (4.24) can be obtained. In this study, we mainly adopt the procedure from [Li, 1997; Sommese & Wampler, 2005]. The corresponding solution from polynomial homotopy continuation technique can be found in APPENDIX D.

4.2.4 Typical hard-spring bistability scenario

By the analytical technique explained in previous section, all the isolated solutions and the corresponding improved corrections of the harmonic balance equation (4.23) can be obtained. In this section, the steady state response with hard-spring bistability is discussed by constructing the relation of the steady state amplitude versus the excitation frequency in Eq. (4.12). The system parameters $\omega_1 = 1$, $\omega_2 = 1$, $k_1 = 0.1$, $k_2 = 0.5$, $k_3 = -0.25$, $k_4 = 0.1$, $k_5 = 0.1$, $k_6 = 0.1$, $k_7 = 0.1$, $\xi_1 = 0.1$, $\xi_2 = 0.2$, $\xi_3 = 0.2$, $\xi_4 = 0.2$, $f_1 = 0.6$, $f_2 = 0.1$, $\lambda = 0.5$ are taken so that the response predominating at the same frequency as the excitation so that the other Fourier components can be neglected in comparison to the bias term and the fundamental harmonic term. When the fractional order $\lambda = 0.5$, the hard-spring bistability scenarios of main system and absorber are illustrated and compared with the results of direct numerical integration of the equations of motion. In Figure 4.13, the solid and broken lines denote stable and unstable branches respectively, dots are results from numerical integration. From Figure 4.13, it is observed that a typical hard-spring bistability scenario occurs. The upper turning point $P(\omega \approx 1.01257)$ is located at the high frequency side and the hysteresis cycle is moving clockwise. More specifically, the response amplitude curve is divided into three regions (I, II and III) with increasing excitation frequency from 0.8 to 1.1. In region I, there is only one stable solution and the response amplitude keeps increasing monotonically for main system and absorber. When the excitation frequency increases to turning point $Q(\omega \approx 0.910265)$, a saddle-node bifurcation occurs and there are three coexisting solutions in region II. The two outer branches are stable while the inner one is unstable. A jump-down occurs with increasing excitation frequency. When the excitation frequency increases to 1.01257, a new fold bifurcation occurs, and there is only one stable solution in region III, the lower branch decreases monotonically with the increasing of excitation frequency from 0.910265. In addition, it can be observed that the obtained harmonic solutions are in excellent agreement with numerical ones.

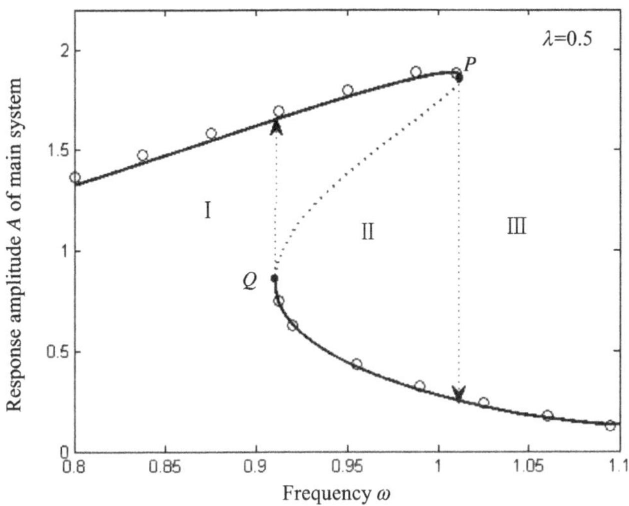

Figure 4.13a Hard-spring bistability scenario of main system for fractional order $\lambda = 0.5$

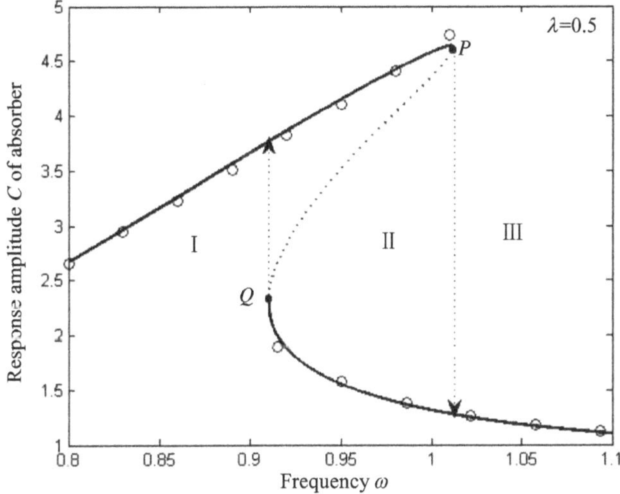

Figure 4.13b Hard-spring bistability scenario of absorber for fractional order $\lambda = 0.5$

4.2.5 Effects of the system parameters on the frequency amplitude curves

In this section, the system parameters are chosen as same as the above section. The effects of fractional order, linear and nonlinear damping coefficients are examined and illustrated by the frequency amplitude curves.

4.2.5.1 Effect of fractional order

Figures 4.14 and 4.15 are presented to illustrate the effects of the fractional order on the frequency response curve. The steady state approximate solution is assumed to be Eq. (4.18). The corresponding frequency amplitude curves are shown in Figures 4.14 and 4.15. The solid lines and broken lines correspond to stable and unstable states, respectively. Figure 4.14 illustrates the frequency versus response amplitude curves for fractional order $\lambda = 0.2, 0.5$ and 0.8 respectively. From Figure 4.14, we can see (i) when excitation frequency increases from 0.5 to turning points $Q_1(\omega \approx 0.89677)$, $Q_2(\omega \approx 0.910265)$ and $Q_3(\omega \approx 0.91782)$ for fractional orders $\lambda = 0.2, 0.5$ and 0.8 respectively, only one stable harmonic solution exists and the response amplitude increases monotonously. (ii) The response amplitudes appear hysteresis phenomena which take place at excitation frequency is in between turning points Q_1 to $P_1(\omega \approx 0.948858)$, Q_2 to $P_2(\omega \approx 1.01257)$, and Q_3 to $P_3(\omega \approx 1.1887)$ to respectively. Although the response curves are roughly of similar shape, the curvatures at the turning points are different for the above cases. (iii) it is observed that there are the considerable differences for different fractional order systems. That is, as the increasing of fractional order from 0.2 to 0.8, the bigger and bigger region of hard-spring bistability occurs, and the peak amplitude of steady state response becomes bigger and bigger. Figure 4.15 illustrates the frequency versus response amplitude curves for excitation frequency increases from 1.2 to 2 corresponding to fractional order $\lambda = 0.1, 0.2, 0.3, 0.4$ and 0.5 respectively. It can be observed that (i) all the response amplitudes of main system keep increasing at first, then decreasing and only one stable solution exist. (ii) The peak amplitude points of forcing frequency for main system keep increasing in the order of the fractional order $\lambda = 0.1, 0.2, 0.3, 0.4$ and 0.5. (iii) The response amplitude of absorber keeps decreasing all the way. To further verify these predictions analytically, they are compared with the numerical simulation in Figures 4.16 and 4.17 at point $\omega=1$ for fractional order $\lambda=0.5$ and 0.2 respectively. The solid lines and starts denote the numerical solutions and stable analytical ones, respectively. The Fourier coefficients of the corresponding analytical solutions are tabulated in Table 4.4. The present analytical predictions match well with the numerical integration solutions.

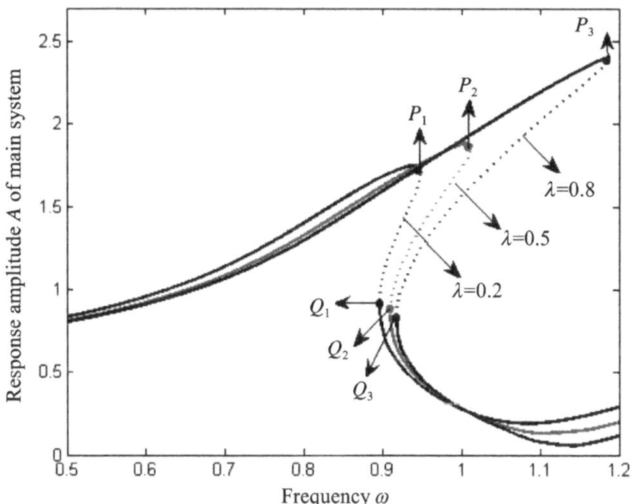

Figure 4.14a Steady state frequency amplitude curve of main system for fractional order λ=0.2,0.5,0.8

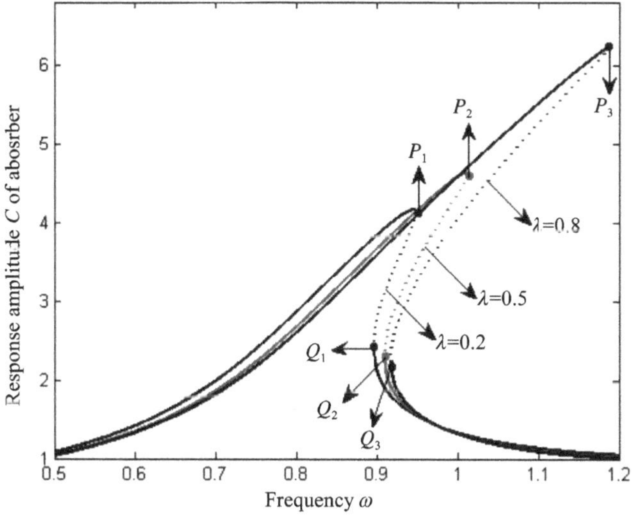

Figure 4.14b Steady state frequency amplitude curve of absorber for fractional order λ=0.2,0.5,0.8

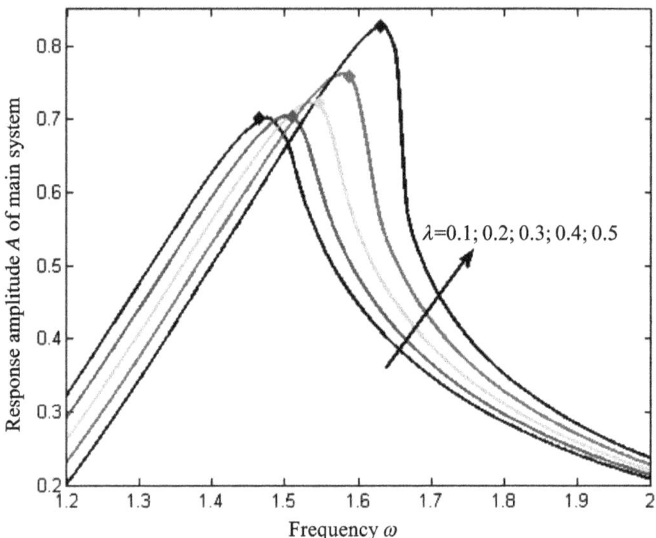

Figure 4.15a Response amplitude curve of main system for fractional order

λ=0.1,0.2,0.3,0.4,0.5

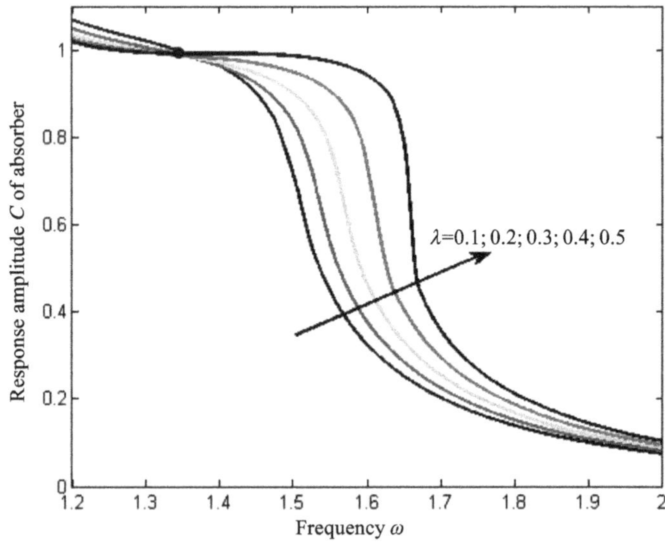

Figure 4.15b Response amplitude curve of absorber for fractional order

λ=0.1,0.2,0.3,0.4,0.5

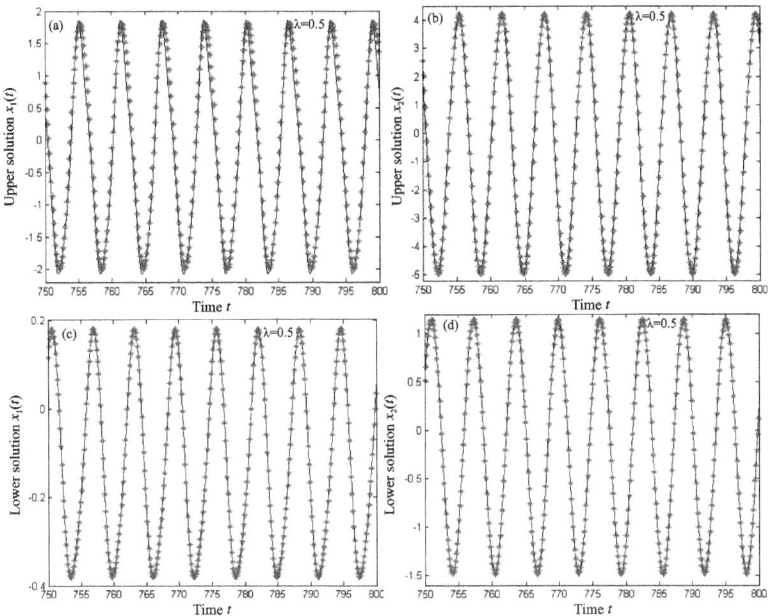

Figure 4.16 Comparisons of the present analytical solutions with numerical results for fractional order λ=0.5. (a) Upper solution $x_1(t)$, (b) Upper solution $x_2(t)$, (c) Lower solution $x_1(t)$, (d) Lower solution $x_2(t)$

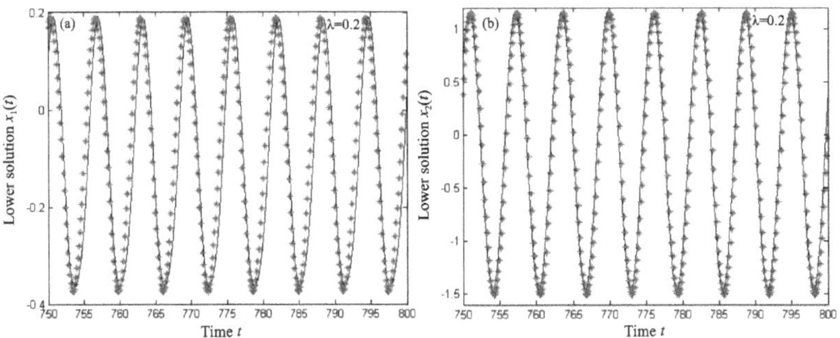

Figure 4.17 Comparisons of the present analytical solutions with numerical results for fractional order λ=0.2. (a) Lower solution $x_1(t)$, (b) Lower solution $x_2(t)$

Table 4.4 The Fourier coefficients of stable responses with ω=1 and fractional order λ=0.5, λ=0.2

Solutions	$\lambda = 0.5$				$\lambda = 0.2$	
	Upper solution $x_1(t)$	Upper solution $x_2(t)$	Lower solution $x_1(t)$	Lower solution $x_2(t)$	Lower solution $x_1(t)$	Lower solution $x_2(t)$
Constant	-0.09743	-0.38847	-0.11882	-0.16560	-0.09310	-0.18460
$\cos(\omega t)$	0.46861	0.73044	-0.26272	-1.28890	-0.24483	-1.27790
$\sin(\omega t)$	1.80850	4.52431	0.08286	-0.20382	0.12813	-0.34360
$\cos(2\omega t)$	0.00651	-0.03276	0.02937	-0.00837	0.00265	-0.00545
$\sin(2\omega t)$	0.01905	-0.03567	0.00467	0.01768	0.00136	0.01855

continued

Solutions	$\lambda=0.5$				$\lambda=0.2$	
	Upper solution $x_1(t)$	Upper solution $x_2(t)$	Lower solution $x_1(t)$	Lower solution $x_2(t)$	Lower solution $x_1(t)$	Lower solution $x_2(t)$
$\cos(3\omega t)$	-0.02811	0.02179	0.00384	-0.00408	0.00254	-0.00244
$\sin(3\omega t)$	-0.01411	-0.00999	0.00344	-0.00320	0.00518	-0.00504

4.2.5.2 Effect of the linear damping

The subsection presents the effects of the linear damping coefficient on periodic motion of the model system with fractional derivative of order one-half. The steady state amplitudes versus frequency responses under linear damping coefficient $\xi_1=-0.1,0.1,0.2,0.5$ and 1 are shown in Figure 4.18. From Figure 4.18, it can be observed that (i) when the linear damping coefficient $\xi_1=-0.1,0.1,0.2,0.5$, the system response depicts smaller and smaller region of hard-spring bistability. There is a similar dynamics to that described in Figure 4.14. Moreover, two stable solutions coexist when the forcing frequency increases from turning points $Q_1(\omega\approx0.8938)$ to $P_1(\omega\approx1.0855)$, $Q_2(\omega\approx0.910265)$ to $P_2(\omega\approx1.01257)$, $Q_3(\omega\approx0.916422)$ to $P_3(\omega\approx0.9831)$, and $Q_4(\omega\approx0.9254)$ to $P_4(\omega\approx0.9274)$ corresponding to linear damping coefficient $\xi_1=-0.1,0.1,0.2,0.5$, respectively. (ii) When the linear damping coefficient $\xi_1=1$ all the response amplitudes keep increasing at first, then decreasing. As a result, it can be concluded that different linear dampings not only can change the amplitude peak of the system but also eliminate the saddle-node bifurcation. Figures 4.19 and 4.20 give the comparisons of the present analytical solutions with numerical results for linear damping coefficient $\xi_1=-0.1$ and $\xi_1=0.5$, respectively. The solid lines and starts denote the numerical solutions and stable analytical ones, respectively. The Fourier coefficients of the corresponding analytical solutions are tabulated in Table 4.5.

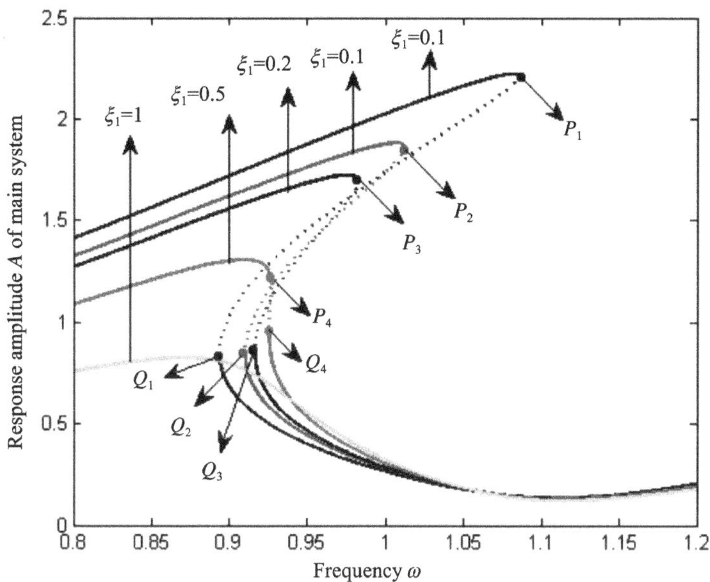

Figure 4.18a Response amplitude curve of main system under different linear damping coefficients with fractional derivative of order one-half

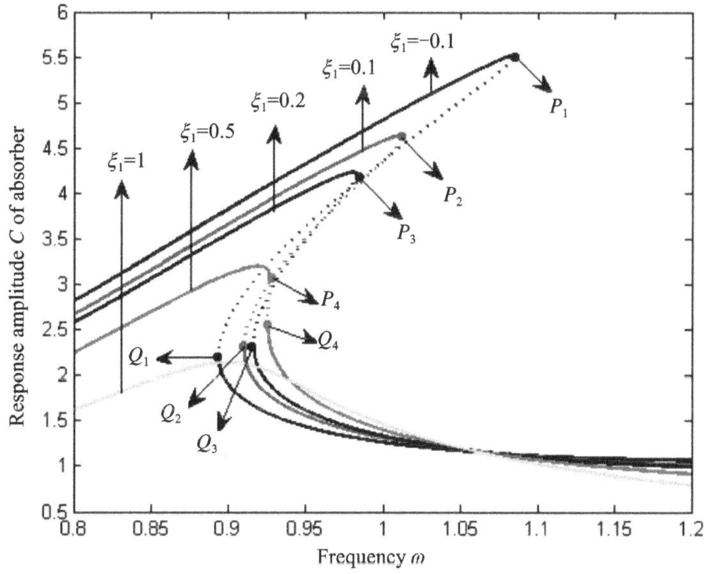

Figure 4.18b Response amplitude curve of absorber under different linear damping coefficients with fractional derivative of order one-half

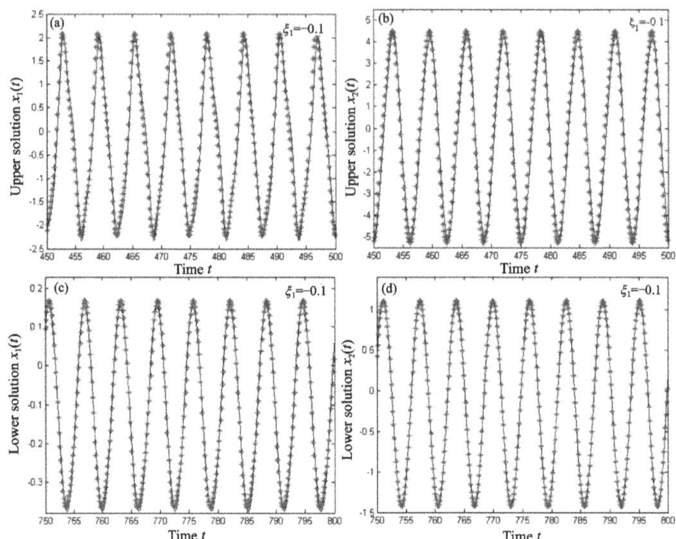

Figure 4.19 Comparisons of the present analytical solutions with numerical results for fractional order $\xi_1 = -0.1$. (a) Upper solution $x_1(t)$, (b) Upper solution $x_2(t)$, (c) Lower solution $x_1(t)$, (d) Lower solution $x_2(t)$

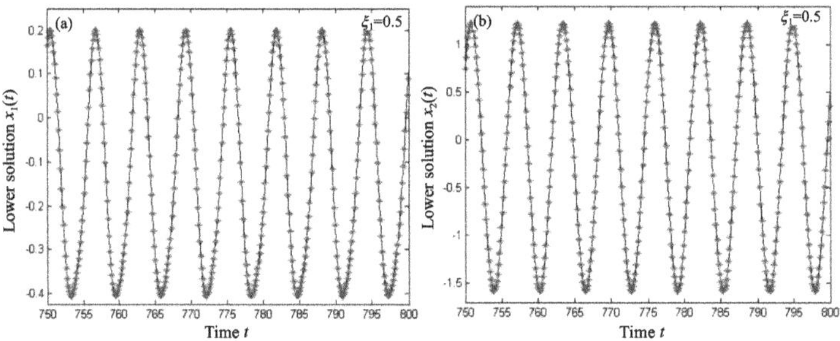

Figure 4.20 Comparisons of the present analytical solutions with numerical results for fractional order $\xi_1 = 0.5$. (a) Lower solution $x_1(t)$, (b) Lower solution $x_2(t)$

Table 4.5 The Fourier coefficients of stable responses with $\omega=1$ and linear damping coefficient $\xi_1 = -0.1$ and $\xi_1 = 0.5$

Solutions	$\xi_1 = -0.1$				$\xi_1 = 0.5$	
	Upper solution $x_1(t)$	Upper solution $x_2(t)$	Lower solution $x_1(t)$	Lower solution $x_2(t)$	Lower solution $x_1(t)$	Lower solution $x_2(t)$
Constant	-0.23523	-0.41501	-0.11177	-0.15617	-0.13098	-0.18242
$\cos(\omega t)$	1.64572	3.80473	-0.25552	-1.22209	-0.26656	-1.38922
$\sin(\omega t)$	1.11416	2.83420	0.06051	-0.29848	0.13663	0.04021
$\cos(2\omega t)$	0.14407	-0.32375	0.02341	-0.00849	0.03332	0.00101
$\sin(2\omega t)$	0.09113	0.07045	0.01022	0.01271	-0.01547	-0.01643
$\cos(3\omega t)$	0.16832	-0.15518	0.00235	-0.00283	0.00628	-0.00615
$\sin(3\omega t)$	0.14763	0.03099	0.00369	-0.00374	0.00009	-0.00054

4.2.5.3 Effect of the nonlinear damping

In this subsection, to study the effect of the nonlinear damping coefficient on periodic motion of the model system (4.12) with fractional derivative of order one-half, eight examples of $\xi_2 = -0.2, 0.1, 0.2, 0.5, 2, 5, 10, 15$ are given. From Figures 4.21(a) and 4.21 (b), it is shown that (i) the response curves for these four cases of $\xi_2 = -0.2, 0.1, 0.2, 0.5$ have two stable branches and one unstable branch separated by two saddle node points. (ii) Although the response curves of different nonlinear damping coefficients are roughly in similar shape, the curvatures and turning points are different. These figures indicate that the hysteresis phenomena occur in the range of forcing frequency delimited between $Q_1(\omega \approx 0.88775)$ and $P_1(\omega \approx 1.145)$ for $\xi_2 = -0.2$; $Q_2(\omega \approx 0.90563)$ and $P_2(\omega \approx 1.0435)$ for $\xi_2 = 0.1$; $Q_3(\omega \approx 0.910265)$ and $P_3(\omega \approx 1.01257)$ for $\xi_2 = 0.2$; $Q_4(\omega \approx 0.9218)$ and $P_4(\omega \approx 0.9811)$ for $\xi_2 = 0.5$, respectively. (iii) With increasing of nonlinear damping coefficient from -0.2 to 0.5, the corresponded frequency of upper turning point gradually becomes large. (iv) When the nonlinear damping coefficient $\xi_2 = 2, 5, 10$, the system response still depicts smaller and smaller region of hard-spring bistability, the corresponding two stable solutions coexist when the forcing frequency increases from turning points $Q_5(\omega \approx 0.955261)$ to $P_5(\omega \approx 0.975714)$, $Q_6(\omega \approx 0.98546)$ to $P_6(\omega \approx 0.99419)$, $Q_7(\omega \approx 1.00903)$ to $P_7(\omega \approx 1.012178)$, respectively. However, the corresponded frequency of upper turning point gradually become small when the nonlinear damping coefficient from 2 to 10. (v) When the nonlinear damping coefficient $\xi_2 = 15$, all the response amplitudes keep increasing at first, then decreasing. As a result, it can be concluded that different nonlinear damping coefficients not only can change the amplitude peak of the system but also eliminate the saddle-node bifurcation. Figures 4.23–4.26 give the comparisons of the present analytical solutions with numerical results for nonlinear damping coefficient $\xi_2 = -0.2, 0.5, 5$ and 15, respectively. The solid lines and starts denote the numerical solutions and stable analytical ones, respectively. The Fourier coefficients of the corresponding analytical solutions are tabulated in Table 4.6. It has been shown that the present analytical predictions match well with the numerical integration solutions.

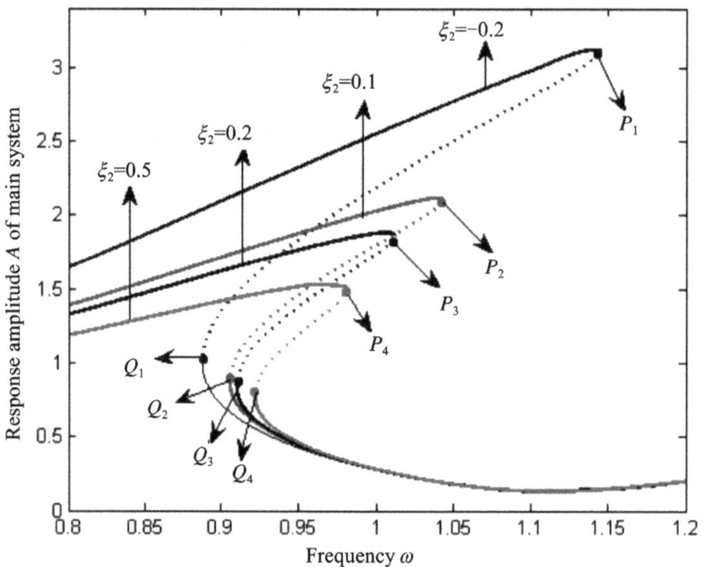

Figure 4.21a Response amplitude curve of main system under different nonlinear damping coefficients $\xi_2=-0.2, 0.1, 0.2$ and 0.5

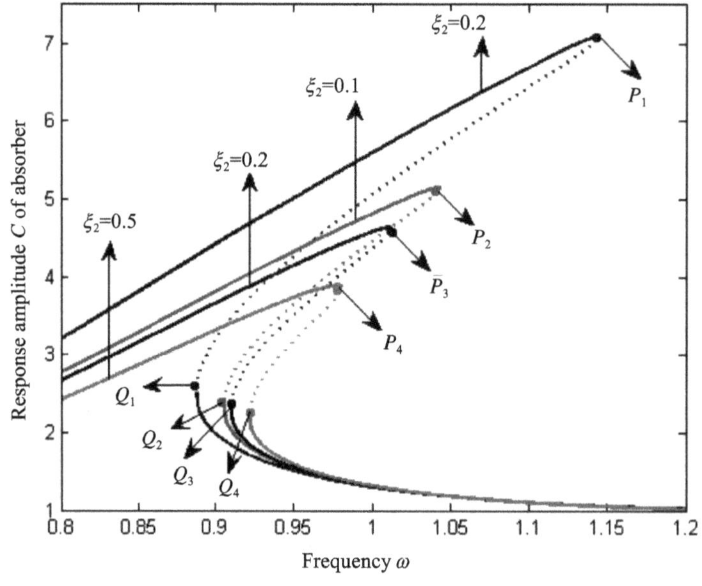

Figure 4.21b Response amplitude curve of absorber under different nonlinear damping coefficients $\xi_2=-0.2, 0.1, 0.2$ and 0.5

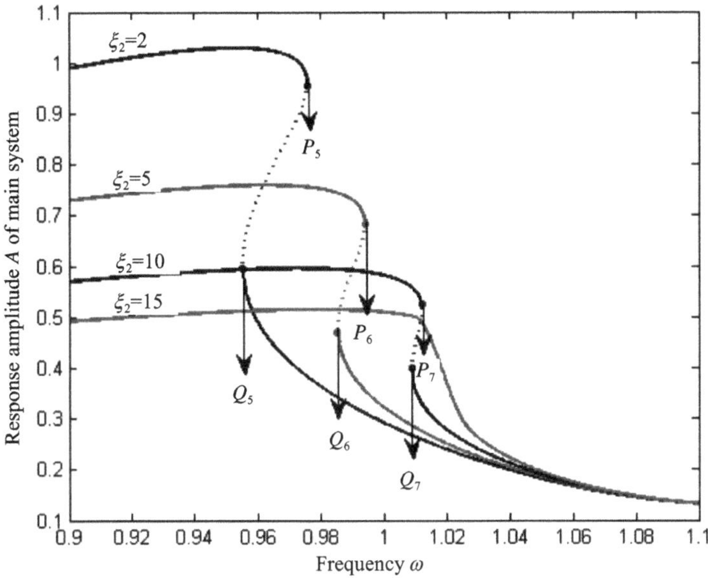

Figure 4.21c Response amplitude curve of main system under different nonlinear damping coefficients $\xi_2 = 2,5,10$ and 15

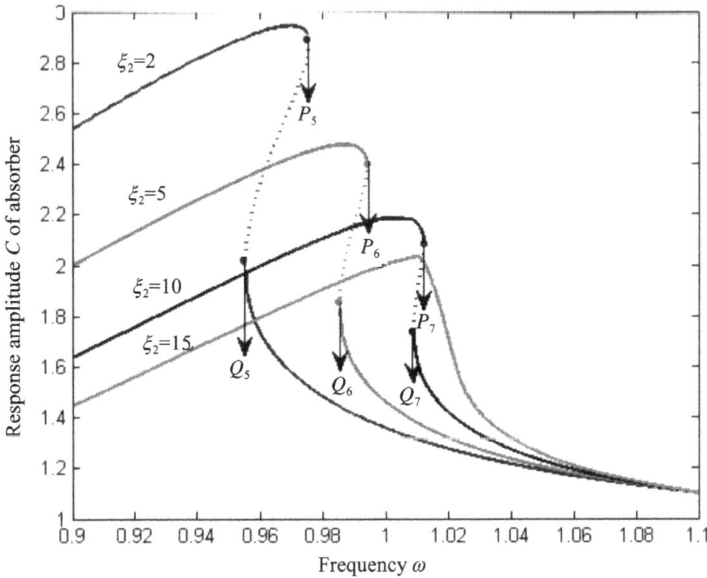

Figure 4.21d Response amplitude curve of absorber under different nonlinear damping coefficients $\xi_2 = 2,5,10$ and 15

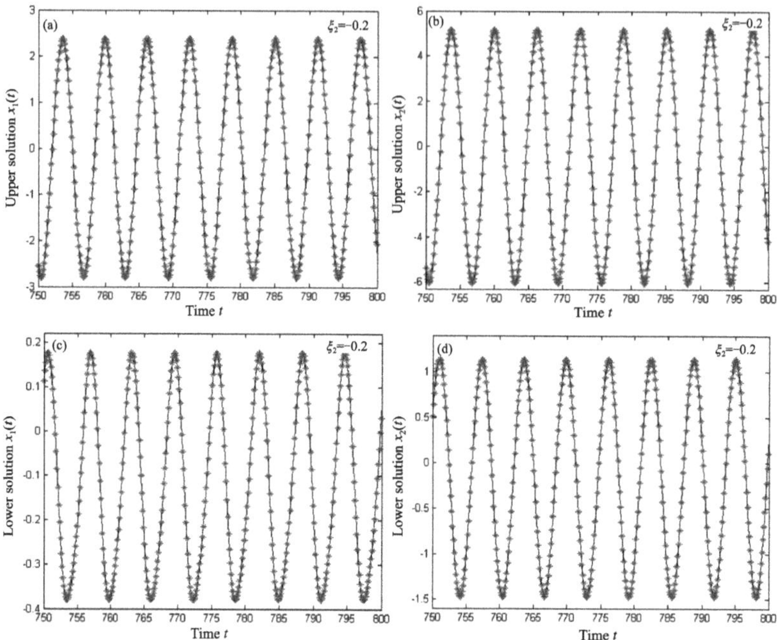

Figure 4.22 Comparisons of the present analytical solutions with numerical results for fractional order $\xi_2 = -0.2$. (a) Upper solution $x_1(t)$, (b) Upper solution $x_2(t)$, (c) Lower solution $x_1(t)$, (d) Lower solution $x_2(t)$

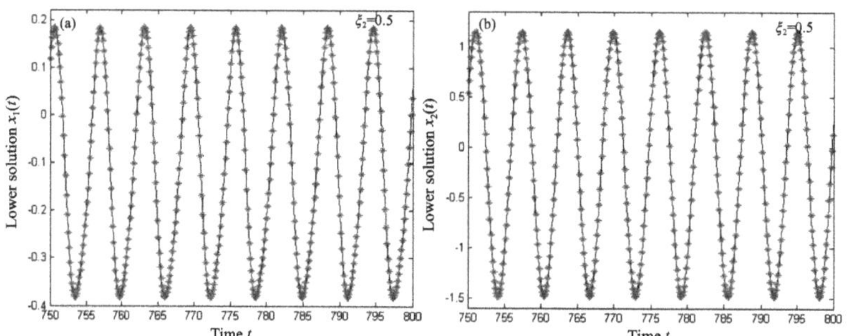

Figure 4.23 Comparisons of the present analytical solutions with numerical results for fractional order $\xi_2 = 0.5$. (a) Lower solution $x_1(t)$, (b) Lower solution $x_2(t)$

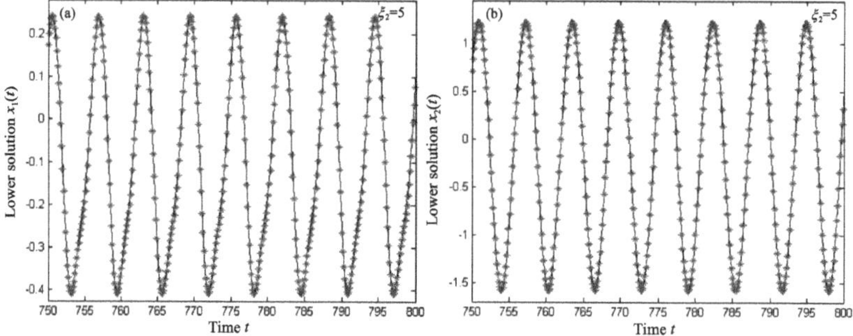

Figure 4.24 Comparisons of the present analytical solutions with numerical results for fractional order $\xi_2 = 5$. (a) Lower solution $x_1(t)$, (b) Lower solution $x_2(t)$

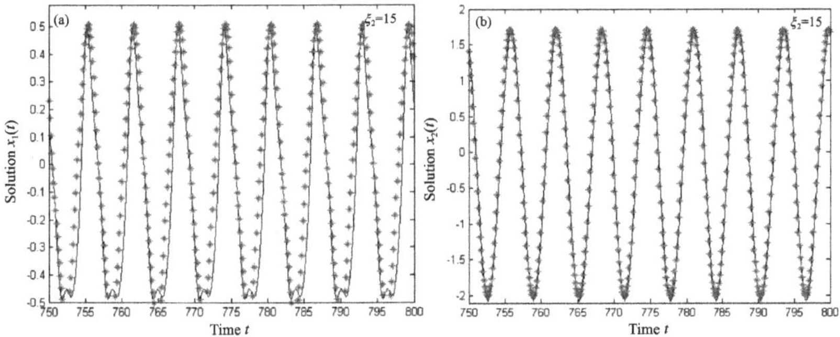

Figure 4.25 Comparisons of the present analytical solutions with numerical results for fractional order $\xi_2 = 15$. (a) Solution $x_1(t)$, (b) Solution $x_2(t)$

Table 4.6a. The Fourier coefficients of stable responses with $\omega = 1$ and nonlinear damping coefficient $\xi_2 = -0.2$ and $\xi_2 = 0.5$

Solutions	$\xi_2 = -0.2$				$\xi_2 = 0.5$	
	Upper solution $x_1(t)$	Upper solution $x_2(t)$	Lower solution $x_1(t)$	Lower solution $x_2(t)$	Lower solution $x_1(t)$	Lower solution $x_2(t)$
Constant	-0.22578	-0.41258	-0.12031	-0.16704	-0.11769	-0.16456
$\cos(\omega t)$	2.35294	5.25772	-0.26125	-1.28035	-0.26383	-1.29532
$\sin(\omega t)$	-1.03589	-1.94754	0.08036	-0.21335	0.06479	-0.19641
$\cos(2\omega t)$	0.01913	0.00953	0.02812	-0.00729	0.03041	-0.00925
$\sin(2\omega t)$	-0.00502	0.02563	0.00382	0.01797	0.00528	0.01751
$\cos(3\omega t)$	0.02220	0.01350	0.00353	-0.00387	0.00410	-0.00424
$\sin(3\omega t)$	-0.01289	0.01084	0.00307	-0.00325	0.00336	-0.00316

Table 4.6b The Fourier coefficients of stable responses with $\omega = 1$ and nonlinear damping coefficient $\xi_2 = 5$ and $\xi_2 = 15$

Solutions	$\xi_2 = 5$		$\xi_2 = 15$	
	Lower solution $x_1(t)$	Lower solution $x_2(t)$	Solution $x_1(t)$	Solution $x_2(t)$
Constant	-0.10611	-0.15251	-0.04021	-0.17165
$\cos(\omega t)$	-0.27945	-1.39093	0.09869	-0.26988
$\sin(\omega t)$	0.12519	-0.04143	0.49972	1.83658
$\cos(2\omega t)$	0.06089	-0.02980	-0.00539	-0.02480
$\sin(2\omega t)$	0.00350	0.02889	0.01519	-0.00524
$\cos(3\omega t)$	0.01141	-0.00713	-0.00222	0.01840
$\sin(3\omega t)$	0.00269	-0.00095	-0.03145	0.01021

4.2.6 Concluding Remarks

The chapter focuses on the steady state bifurcation of a two-degree-of-freedom vibration system including quadratic and cubic nonlinearities subjected to external and parametric excitation forces under a nonlinear absorber with damping modelled by a fractional derivative. The system is governed by the fractional order differential equations. The effects of fractional derivative order, linear and nonlinear damping coefficients on the nonlinear behavior have been illustrated since fractional order and system damping are often used as a control parameter to achieve certain motions.

The nonlinear equations are solved analytically by a coupled method of harmonic balance and polynomial homotopy continuation algorithm to investigate the hard-spring bistability scenario and effect of system parameters on the system behavior. Based on the obtained steady state responses, it is found that different damping coefficients not only change the amplitude peak of the system but also can eliminate the saddle-node bifurcation. Moreover, the system response depicts bigger and bigger region of hard-spring bistability with increasing of fractional order, however, the system response depicts smaller and smaller region of hard-spring bistability as the increasing of linear and nonlinear damping coefficients. In a more systematic way, these interesting results are illustrated in Table 4.7.

Table 4.7 The effects of the fractional order, linear and nonlinear damping coefficients on periodic motion of the model system

Effects of system parameters	with increasing of fractional order $\lambda, 0 < \lambda \le 1$	with increasing of linear damping coefficient ξ_1	with increasing of nonlinear damping coefficient ξ_2
Region of hard-spring bistability	↗	↙	↙
Amplitude peak of main system and absorber	↗	↙	↙
Critical parameter for eliminating saddle-node bifurcation		0.53 or so	14.5 or so

In addition, the presented approximate predictions corresponding to steady state resonance responses for the given points are illustrated and compared, which they are in good agreement with numerical integration ones. The above results have potential applications on the other fractional order systems.

Chapter 5 Applications of Fractional Derivative to Structural Engineering

5.1 Resonance responses and bifurcation of viscoelastic arch system

5.1.1 Introduction

Arches are common structural members appearing in civil and mechanical engineering as sub-elements of more complex structures. A large amount of research activities associated with static and dynamic behaviors of the arches have been conducted during the past few decades. Among which, the nonlinear dynamic responses of the harmonically forced arch structures, e.g. the primary and subharmonic resonance as well the routes to chaos are of special interest to many research communities [Plaut & Hsieh, 1985; Mook et al., 1986; Leung & Fung, 1990; Tien, Namachchivaya & Bajaj, 1994; Tien, Namachchivaya & Malhotra, 1994; Malhotra & Namachchivaya, 1997; Bi & Dai, 2000; Chen, J.S. & Yang, 2007]. Noting that, all above-mentioned researchers almost assumed the arches under their consideration to be elastic and did not take into account any dissipative mechanisms of materials. Therefore, in this section, an extension of fractional calculus models to the nonlinear range of viscoelasticity is attempted [Li, G.G., 2002; Rossikhin, Y.A. & Shitikova, M.V., 2010].

5.1.2 The Equation of Motion for Viscoelastic Arch System

Consider a viscoelastic shallow arch having span L and thickness h shown in Figure 5.1.

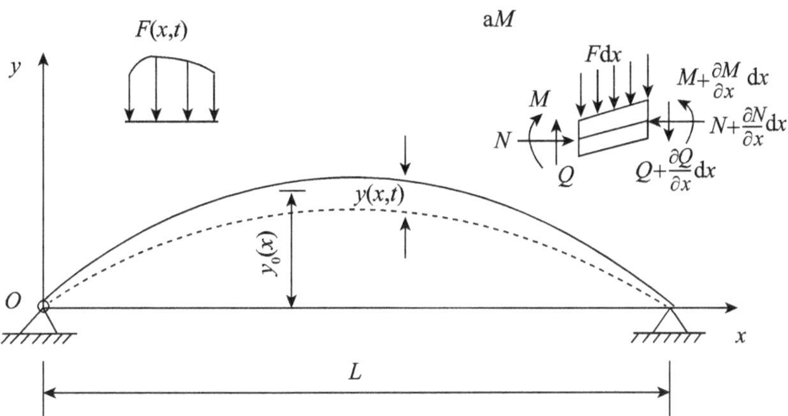

Figure 5.1 Schematic of the viscoelastic shallow arch

The arch has hinged-hinged ends and is subjected to a lateral periodic excitation $F(x,t)$. Let $y_0(x)$ and $y(x,t)$ denote the initial and deformed configurations, respectively. The governing equation can then be written as

$$\frac{\partial Q}{\partial x} = -F(x,t) - \rho A \frac{\partial^2 y}{\partial^2 t},$$

$$\frac{\partial M}{\partial x} = Q - N \frac{\partial y}{\partial x}, \tag{5.1}$$

where ρ is the mass density of the arch, A is the area of cross-section, M is the bending moment, Q is the lateral force, and N is the horizontal force. From Eq. (5.1), we have

$$\frac{\partial^2 M}{\partial x^2} + N \frac{\partial^2 y}{\partial x^2} + \rho A \frac{\partial^2 y}{\partial^2 t} + F(x,t) = 0. \tag{5.2}$$

Assume that the material of the arch obeys the fractional Kelvin-Voigt model

$$\sigma(t) = E_0 \varepsilon(t) + E_1 D_t^\alpha [\varepsilon(t)] = E_0 (1 + \eta D_t^\alpha) \varepsilon(t) \overset{\Delta}{=} S\varepsilon(t), \tag{5.3}$$

where $\eta = E_1 / E_0$ is the material modulus ratio. The bending moment is given by

$$M = I[E_0(\frac{\partial^2 y}{\partial x^2} - \frac{\partial^2 y_0}{\partial x^2}) + E_1 D_t^\alpha(\frac{\partial^2 y}{\partial x^2} - \frac{\partial^2 y_0}{\partial x^2})] = IS(\frac{\partial^2 y}{\partial x^2} - \frac{\partial^2 y_0}{\partial x^2}), \tag{5.4}$$

And the axial resistant force is

$$N = \frac{E_0 A}{2L} \int_0^L [(\frac{\partial y_0}{\partial x})^2 - (\frac{\partial y}{\partial x})^2] dx + \frac{E_1 A}{2L} \int_0^L D_t^\alpha [(\frac{\partial y_0}{\partial x})^2 - (\frac{\partial y}{\partial x})^2] dx$$

$$= \frac{A}{2L} S \int_0^L [(\frac{\partial y_0}{\partial x})^2 - (\frac{\partial y}{\partial x})^2] dx, \tag{5.5}$$

where I is the moment of inertia. Then Eq. can be rewritten as

$$IS(\frac{\partial^4 y}{\partial x^4} - \frac{\partial^4 y_0}{\partial x^4}) + \frac{A}{2L} \frac{\partial^2 y}{\partial x^2} S \int_0^L [(\frac{\partial y_0}{\partial x})^2 - (\frac{\partial y}{\partial x})^2] dx + \rho A \frac{\partial^2 y}{\partial^2 t} + F(x,t) = 0, \tag{5.6}$$

having the boundary conditions

$$y = y_0 = 0, \frac{\partial^2 y}{\partial x^2} = \frac{\partial^2 y_0}{\partial x^2} = 0, \text{ at } x = 0, x = L. \tag{5.7}$$

Substituting $w(x,t) = y(x,t) - y_0(x)$ into Eq. yields

$$IS\frac{\partial^4 w}{\partial x^4} - \frac{A}{2L}(\frac{\partial^2 w}{\partial x^2} + \frac{\partial^2 y_0}{\partial x^2}) S \int_0^L [(\frac{\partial w}{\partial x})^2 + 2\frac{\partial w}{\partial x}\frac{\partial y_0}{\partial x}] dx + \rho A \frac{\partial^2 w}{\partial^2 t} + F(x,t) = 0. \tag{5.8}$$

Next, the Galerkin method is applied to simplify the system from PDE to ODE. To conform to the boundary conditions, the solution of Eq. can be expanded into Fourier series

$$w(x,t) = \sum_{n=1}^{\infty} u_n(t) \sin(\frac{n\pi}{L} x). \tag{5.9}$$

Additionally, the initial configuration of the arch is assumed to be sinusoidal and subjected to sinusoidal distribution loads.

$$F(x,t) = \rho A f \sin(\frac{\pi}{L} x) \cos(\omega t), \quad y_0(x) = y_0 \sin(\frac{\pi}{L} x).$$

Then, the one term Galerkin approximation gives

$$u'' + k_1 u + k_2 u^2 + k_3 u^3 + \eta[k_1 D_t^\alpha u + \frac{1}{3} k_2 D_t^\alpha (u^2) + \frac{2}{3} k_2 u D_t^\alpha u + k_3 u D_t^\alpha (u^2)] + f \cos(\omega t) = 0, \tag{5.10}$$

where u'' denotes the differentiation of u with respect to t twice, f (m/s^2) represents the excitation amplitude, and

$$k_1 = (\frac{\pi}{L})^4 \frac{(2I + A y_0^2) E_0}{2\rho A}, k_2 = (\frac{\pi}{L})^4 \frac{3 y_0 E_0}{4\rho}, k_3 = (\frac{\pi}{L})^4 \frac{E_0}{4\rho}. \tag{5.11}$$

Just for simplicity, the unit of f has been omitted in following discussion and graphical representation. When $\alpha = 1$ the fractional oscillator reduces to a nonlinear oscillator representing an arch with viscous damping

$$u'' + k_1 u + k_2 u^2 + k_3 u^3 + \eta(k_1 + \frac{4}{3}k_2 u + 2k_3 u^2)u' + f\cos(\omega t) = 0. \tag{5.12}$$

In the following section, we will study the steady state fundamental resonance response of the nonlinear fractional order equations and investigate the effects of various parameters on the response.

5.1.3 Solution Procedure for Steady State Responses

Due to a fact that the HB method can generate both the stable and unstable steady state solutions for nonlinear differential equations, the HB method is used for steady state solutions for fractional differential equation (5.10). Eq.(5.12) can be rewritten as follows

$$\omega^2 \ddot{u} + k_1 u + k_2 u^2 + k_3 u^3 + \eta\omega^\alpha[k_1 D_\tau^\alpha u + \frac{1}{3}k_2 D_\tau^\alpha(u^2) + \frac{2}{3}k_2 u D_\tau^\alpha u + k_3 u D_\tau^\alpha(u^2)]$$
$$+ f\cos(\tau) = 0, \tag{5.13}$$

by introducing a new time scale $x = \omega t$. Wherein, $\ddot{u}$ denotes the differentiation of u with respect to τ twice. Introduce a function Φ that represents the whole of the fractional equation (5.13) for the convenience of the following discussion

$$\Phi(\ddot{u}, D_x^\alpha(u^2), D_x^\alpha u, u, C, x)$$
$$\overset{\Delta}{=} f\cos(x) + \omega^2 \ddot{u} + k_1 u + k_2 u^2 + k_3 u^3 \tag{5.14}$$
$$+ \eta\omega^\alpha[k_1 D_x^\alpha u + \frac{1}{3}k_2 D_x^\alpha(u^2) + \frac{2}{3}k_2 u D_x^\alpha u + k_3 u D_x^\alpha(u^2)],$$

where C represents all the constant parameters existing in fractional equation. For the primary resonance, the solution of Eq. (5.13) is assumed to be

$$u = a_0 + a_1\cos(x) + b_1\sin(x). \tag{5.15}$$

Substituting Eq. (5.15) into Eq.(5.13), and equating the coefficients of like harmonic to zero leading to three nonlinear algebraic equations

$$R_0(a_0,a_1,b_1) = \frac{2}{\pi} \int_0^{\pi} \Phi(\ddot{u}, D_x^{\alpha}(u^2), D_x^{\alpha}u, u, C, x)\, \mathrm{d}x = 0,$$

$$R_1^c(a_0,a_1,b_1) = \frac{2}{\pi} \int_0^{\pi} \Phi(\ddot{u}, D_x^{\alpha}(u^2), D_x^{\alpha}u, u, C, x)\cos x\, \mathrm{d}x = 0, \qquad (5.16)$$

$$R_1^s(a_0,a_1,b_1) = \frac{2}{\pi} \int_0^{\pi} \Phi(\ddot{u}, D_x^{\alpha}(u^2), D_x^{\alpha}u, u, C, x)\sin x\, \mathrm{d}x = 0,$$

which are solved by using polynomial homotopy continuation method [Sommese & Wampler, 2005].

5.1.4 Influences of Control Parameters on the Resonance Response

In this subsection, the effects of the control parameters on the steady state primary resonance response are studied in detail by the aforementioned method using three harmonic terms in Eq. (5.15) . In general, the amplitudes of bias term and response are denoted, respectively, as follows:

$$A_0 = |a_0|, \ A_1 = \sqrt{a_1^2 + b_1^2}. \qquad (5.17)$$

The various material and structure parameters of the viscoelastic arch considered here are given by Babouskos & Katsikadelis 2010 as follows:

$$E_0 = 21\mathrm{MPa}, \rho = 7500\mathrm{kg/m}^3, L = 15\mathrm{m}, A = 0.64\mathrm{m}, y_0 = 2, h = 0.48. \qquad (5.18)$$

Then, from Eq. (5.11) and Eq. (5.18) , one has $k_1 = 10.8$, $k_2 = 8.08$, $k_3 = 1.35$.

5.1.4.1 Excitation Amplitude-Response Curves under Different Fractional Orders

To fix the idea, consider the material modulus ratio $\eta = 0.005$ and frequncy $\omega = 3.2$. Figures.5.2–5.8 show the bias term and the amplitudes of the fundamental steady state response with increasing excitation amplitude for the various fractional orders: $\alpha = 0.3$, $\alpha = 0.6$, $\alpha = 0.9$, respectively. In these figures, the solid lines correspond to the stable solutions; while the dashed lines correspond to unstable solutions. A simpler approach is to determine the stability by the Runge-Kutta numerical integration because the initial conditions can be readily supplied by the results of the harmonic balance approximation. When $\alpha = 0.6$, it can be seen from Figure 5.2, the response curve of the system is separated into three parts: the lower, middle and upper regions of the response curves, respectively. The whole schematic are enlarged and presented in Figure. 5.3 and 5.4 for bias term and amplitude of response, respectively. It is easy to find that the lower and upper regions have two stable branches and one unstable branch separated by two saddle node points, while the stable solutions do not exist in middle region. Coexisting of the stable and unstable solutions is responsible for the jump and hysteresis phenomena. In addition, the response curves of the lower and upper regions are roughly in similar shape. However, the curvatures and turning points are different.

The comparisons between analytical results of the primary responses of Eq. (5.15) and the numerical ones are confirmed in Figure. 5.5 using the fourth order Runge-Kutta integration.

To study the influence of the fractional order on the dynamic behaviors, three examples of $\alpha = 0.3, 0.6, 0.9, 1$ are given. The bias term A_0 and the amplitude A_1 defined by Eq. (5.17) are shown in Figure. 5.6 and 5.7 for the lower and upper regions, respectively.

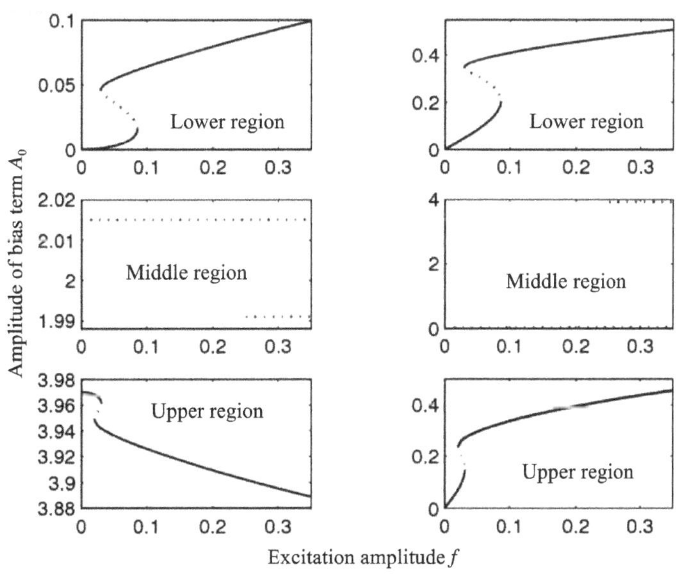

Figure 5.2 Amplitude-response curves of bias terms and for A_0 and A_1 for $\alpha = 0.6$

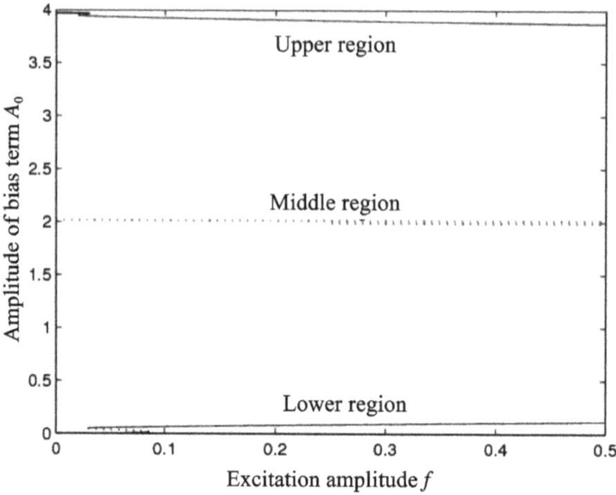

Figure 5.3 Enlargement for excitation amplitude-response curves of bias term A_0 for $\alpha = 0.6$

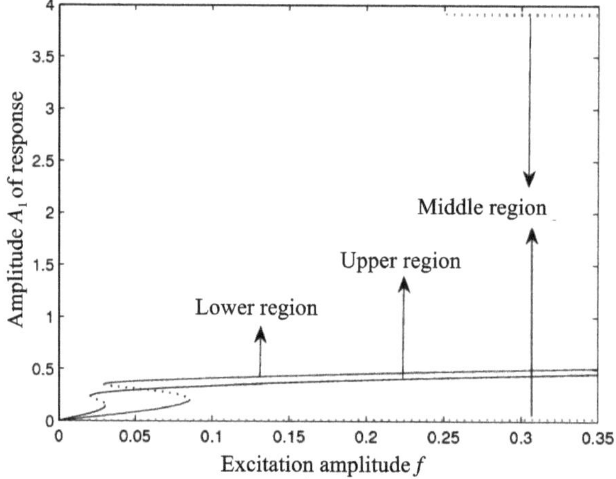

Figure 5.4 Enlargement for amplitude-response curves of amplitude A_1 for $\alpha = 0.6$

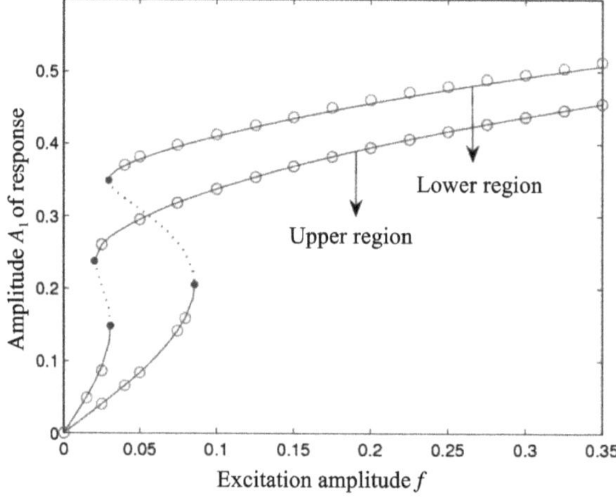

Figure 5.5 Comparisons of the approximately analytical solutions with numerical results

the solid line:stable solution; the dashed line:unstable one; the circles: numerical one

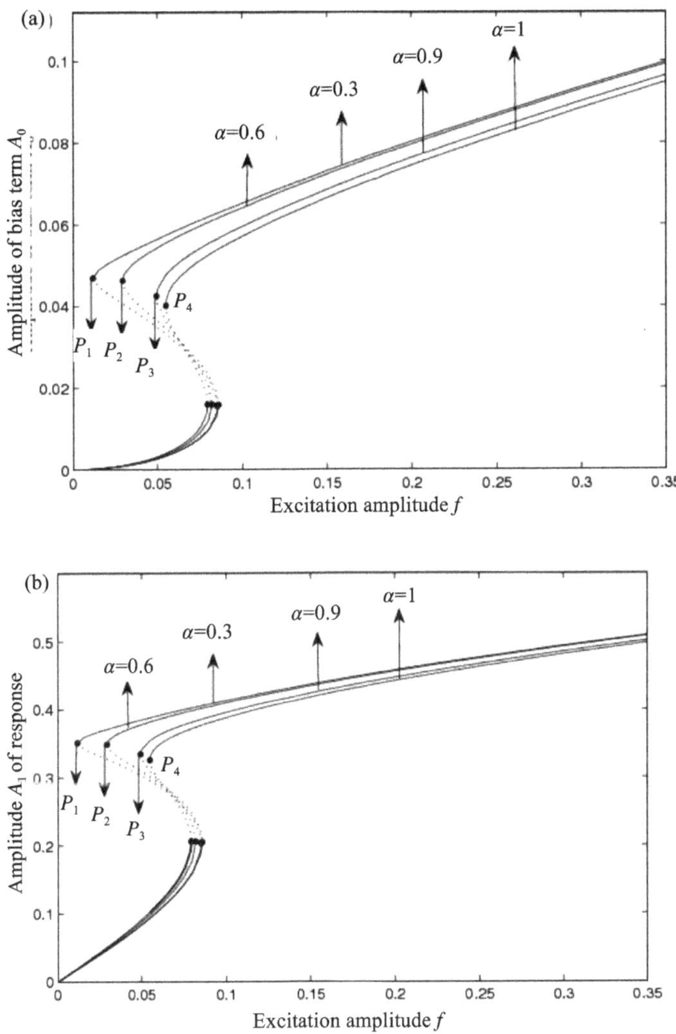

Figure 5.6 Response curves of: (a) amplitude A_0 bias term; (b) amplitude A_1 of response for various fractional orders in lower region

From Figure 5.6 it can be observed that the lower region of response curves for these three cases have two stable branches and one unstable branch separated by two saddle node points. Although the response curves of different fractional orders are roughly in similar shape, the curvatures and turning points are different. The figures indicate that the hysteresis phenomena occur in the range of external forcing amplitude delimited between $f \approx 0.0117085$ and $f \approx 0.085311$ for $\alpha = 0.3$; $f \approx 0.02948$ and $f \approx 0.0859947$ for $\alpha = 0.6$; and $f \approx 0.04945$ and $f \approx 0.082203$ for $\alpha = 0.9$ respectively. For example, from the second picture in Figure 5.6 when

$\alpha = 0.6$, as the external forcing amplitude increases, the amplitude A_1 increases gradually, but then at $f \approx 0.0859947$, the equilibrium solution undergoes a saddle node bifurcation, and the response amplitude jumps up from $A_1 \approx 0.20524572$ to a high amplitude of $A_1 \approx 0.4021323$. Conversely, A_1 decreases gradually with the decrease of f, but then at another bifurcation point P_2 where $f \approx 0.02948$, A_1 drops dramatically from $A_1 \approx 0.34906$ to a low amplitude $A_1 \approx 0.047664$. Moreover, the turning of the amplitude-frequency curve is severe along with the decrease of the fractional order.

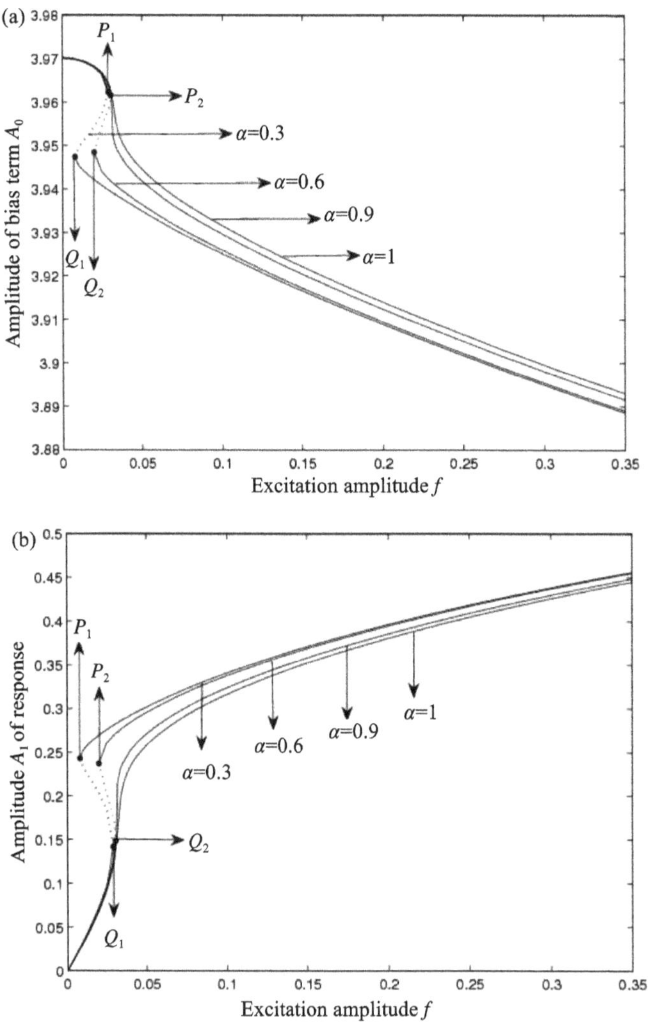

Figure 5.7 Response curves of: (a) amplitude A_0 of bias term; (b) amplitude A_1 of response for various fractional orders in upper region

It is observed from Figure 5.7 that, similar conclusion can be drawn for the upper region of response curves under different fractional orders. It also shows that the smaller the fractional order is, the larger the amplitudes of responses are. In addition, a critical value of fractional order approximately equals to 0.8999985 is noted. When $\alpha \geq 0.8999985$, the bias term A_0 always decreases with the increase of the excitation amplitude and the corresponding amplitude A_1 increases on the same interval. That is, the fractional order damping not only changes the bifurcation curves but also eliminates the saddle nodes on the resonance response, which indicated by Figure 5.7 when $\alpha = 0.9$ and 1. In other words, the damping effect plays a distinct role with the increase of the fractional order.

5.1.4.2 Excitation Frequency-Response Curves under Different Fractional Orders

In this subsection, the response curves of the bias term A_0 and the amplitude A_1 versus circular frequency ω of excitation under different fractional orders are considered, where the material parameter and excitation amplitude are taken as $\eta = 0.005, f = 0.02$, respectively. From Figure 5.8 when $\alpha = 0.3$, it can be found that throughout the region of excitation frequency, there are also three different forms of response curves, in which the middle region is always unstable. The whole schematic of responses for three regions is enlarged to Figs. 5.9 and 5.10 for bias term and amplitude, respectively.

One observes that the hysteresis phenomenon occurs with the increase or decrease of the circular frequencies for the lower and upper regions of the response curves. It should be mentioned that, apart from the usual jump phenomenon, there exists another branch approximately at low frequencies, $\omega \leq 2.132112$ for the lower region and $\omega \leq 2.270233$ for the upper region, respectively. When the forcing amplitude p is increased beyond some critical values, this kind of jump merges with the main jump and the response curve opens up as shown in Figure 5.11 when $f = 0.1$.

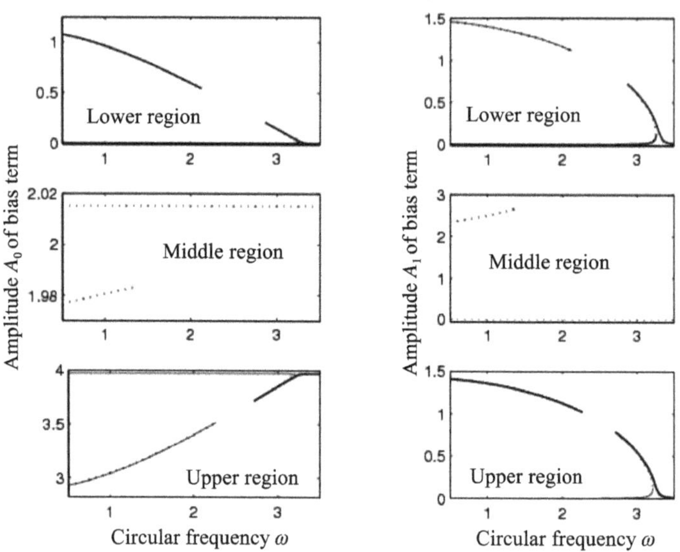

Figure 5.8 Frequency-response curves of bias terms A_0 and amplitude A_1 for $\alpha = 0.3$

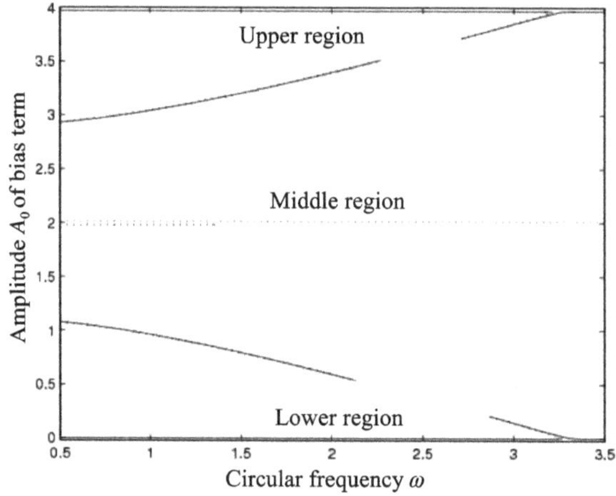

Figure 5.9 Enlargement for frequency-response curves of bias term A_0 for $\alpha = 0.3$

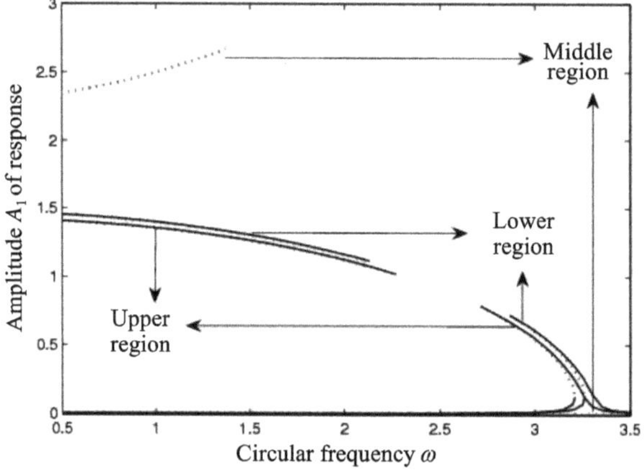

Figure 5.10 Enlargement of frequency-response curves of amplitude A_1 for $\alpha = 0.3$

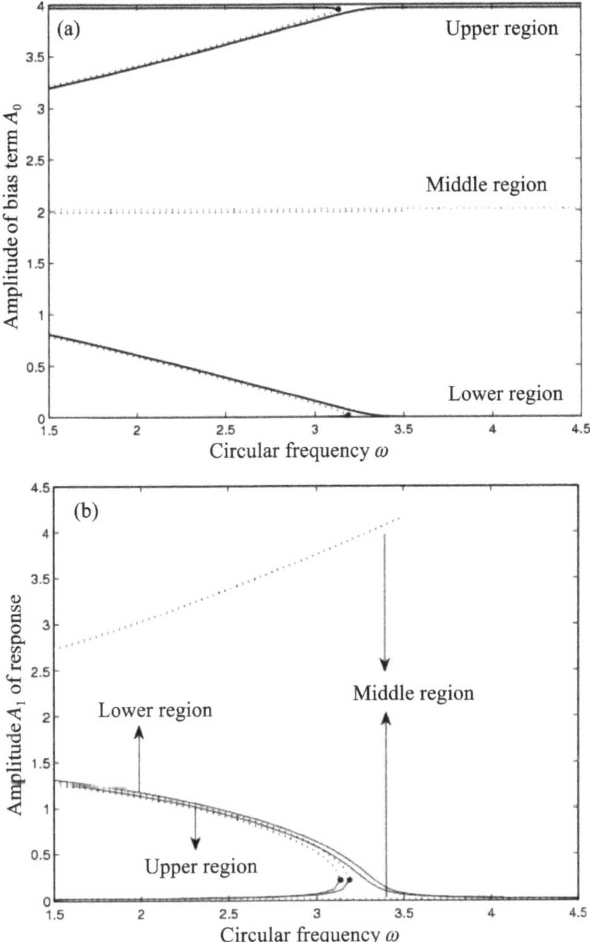

Figure 5.11 Frequency-response curves of: (a) amplitude A_0 of bias term; (b) amplitude A_1 response for $\alpha = 0.3, f = 0.1$

Additionally, the shapes of the response curves of the lower and upper regions are very similar to each other. Therefore, only the lower regions of the response curves for various fractional orders are chosen as numerical examples to show the influences of the fractional order in the subsequent discussion.

The lower regions of the bias term A_0 and the amplitude A_1 for various fractional orders are shown in Figure. 5.12–5.13 respectively. Thee response curves consist of two separated parts which indicate a double jump phenomenon when the forcing frequency is increased. Additionally, with the increase of the fractional order, the hysteresis region shrinks. Specifically, a critical value can be obtained numerically at $\alpha_c \approx 0.7035848186$. When $\alpha > \alpha_c$, the saddle nodes have been eliminated and replaced by a peak point. For example, when $\alpha = 0.9$, the peak points are obtained approximately at $P_3(3.27703, 0.0062958)$ for the bias term A_0 and $P_3(3.27703, 0.12953296)$ for the amplitude A_1, which are depicted by Figs. 5.12(b-1)(b-2) and 5.13(b-1)(b-2).

In summary, from Fig. 5.2–5.13, when the order of the fractional derivative increases, the nonlinear complexity of the system decreases. In other words, with the increase of the fractional order, the effect of viscoelastic damping described by the fractional derivative plays a distinct role in shrinking the region of hysteresis that enhances the stability of the structure.

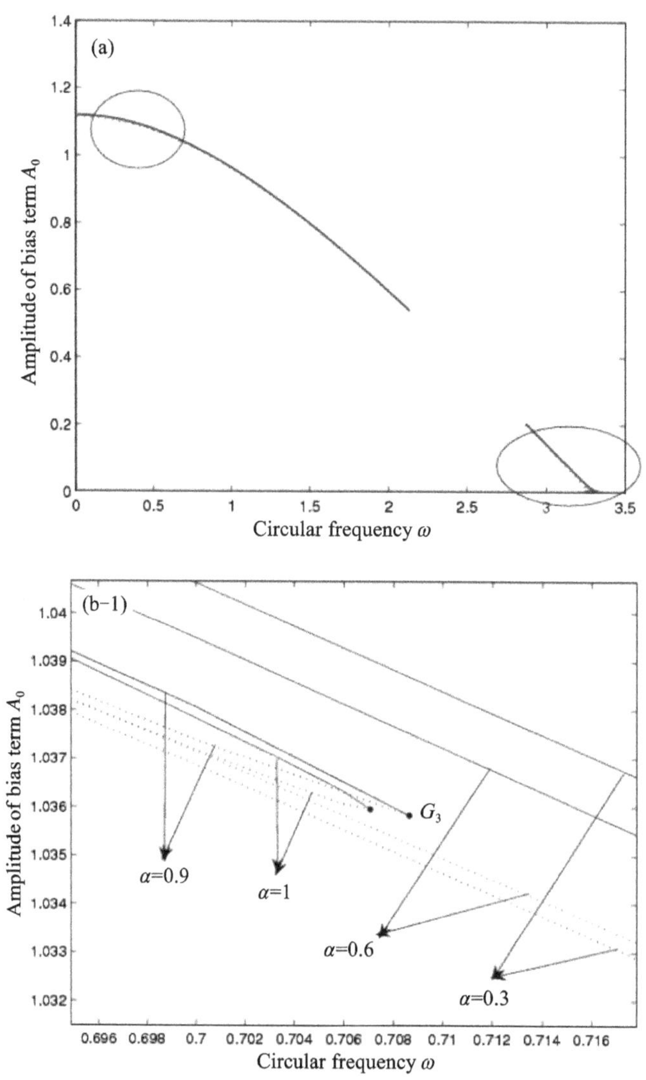

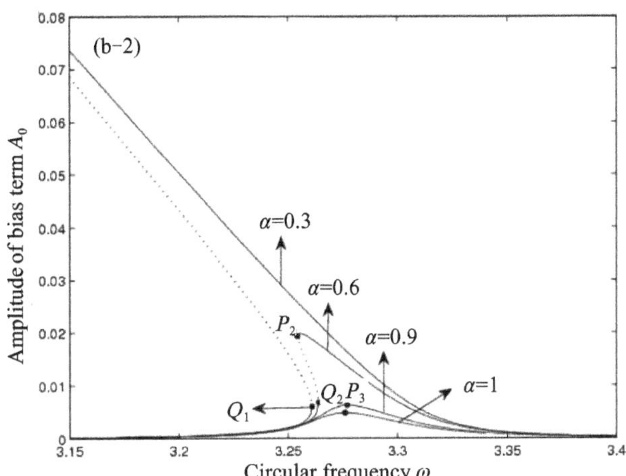

Figure 5.12 Lower region of response curves for amplitude A_0 of bias term for various fractional orders: (a): panoramic view; (b-1)(b-2): local view

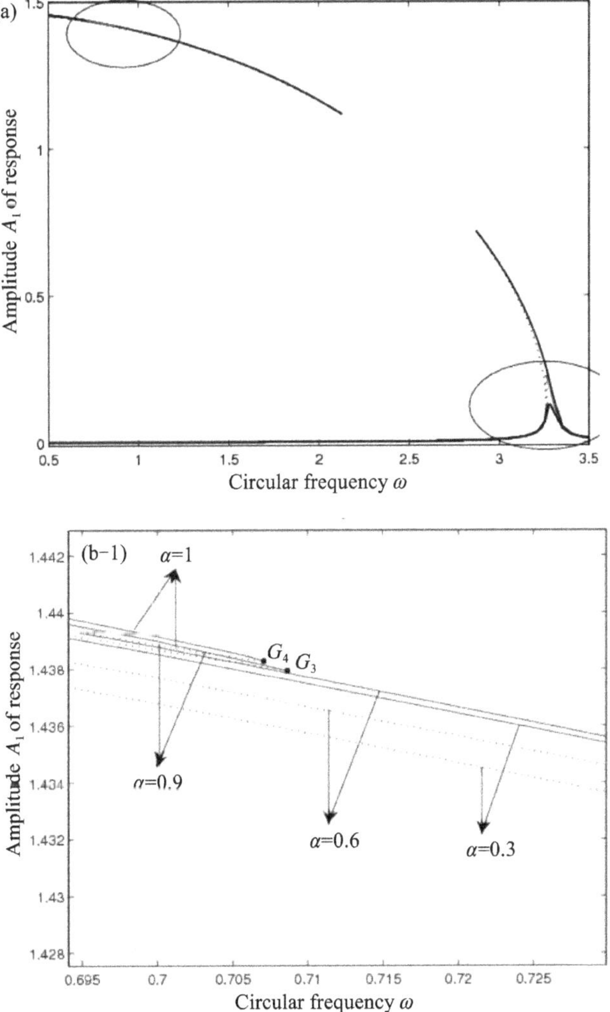

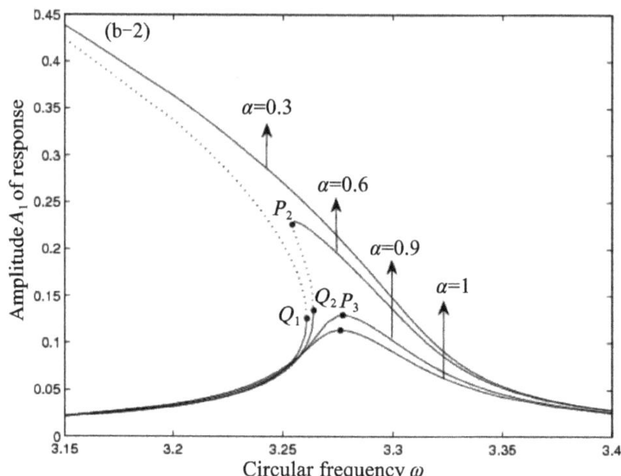

Figure 5.13 Lower region of response curves of amplitude A_1 for various fractional orders: (a): Panoramic view; (b-1)(b-2): local view

5.1.4.3 Influence of Material Parameter

Similarly, the relationships of the bias term and the amplitude versus excitation frequency of the upper region for various material modulus ratio η are shown in Figs. 5.14 and 5.15. The control parameters are defined by $\alpha = 0.5$, $f = 0.02$ and the material modulus ratio is taken as $\eta = 0.002, 0.005, 0.009$ respectively. The solid lines and dashed lines present the stable and unstable solutions respectively.

It can be observed from Fig. 5.14 and 5.15 that the influence of material modulus ratio η on the behavior of the arch is obvious. The jump phenomena occur with the increase or decrease of ω in three cases due to the coexisting of the stable and unstable solutions. Similar to the last section, the hysteresis region becomes smaller with the increase of the material modulus ratio. A conclusion is that, both the fractional order α and the material modulus ratio η play a role of viscoelastic damping that enhances the stability of the structures. A critical value is captured at $\eta_c \approx 0.00829587$, by which the saddle node is eliminated and replaced by a peak one approximately at $P_3(3.234, 3.95943838)$ for the bias term A_0, and $P_3(3.234, 0.166626847)$ for the response amplitude A_1 that are depicted in Figs. 5.14 (b-1)(b-2) and 5.15(b-1)(b-2), respectively, as the material modulus ratio $\eta = 0.009$.

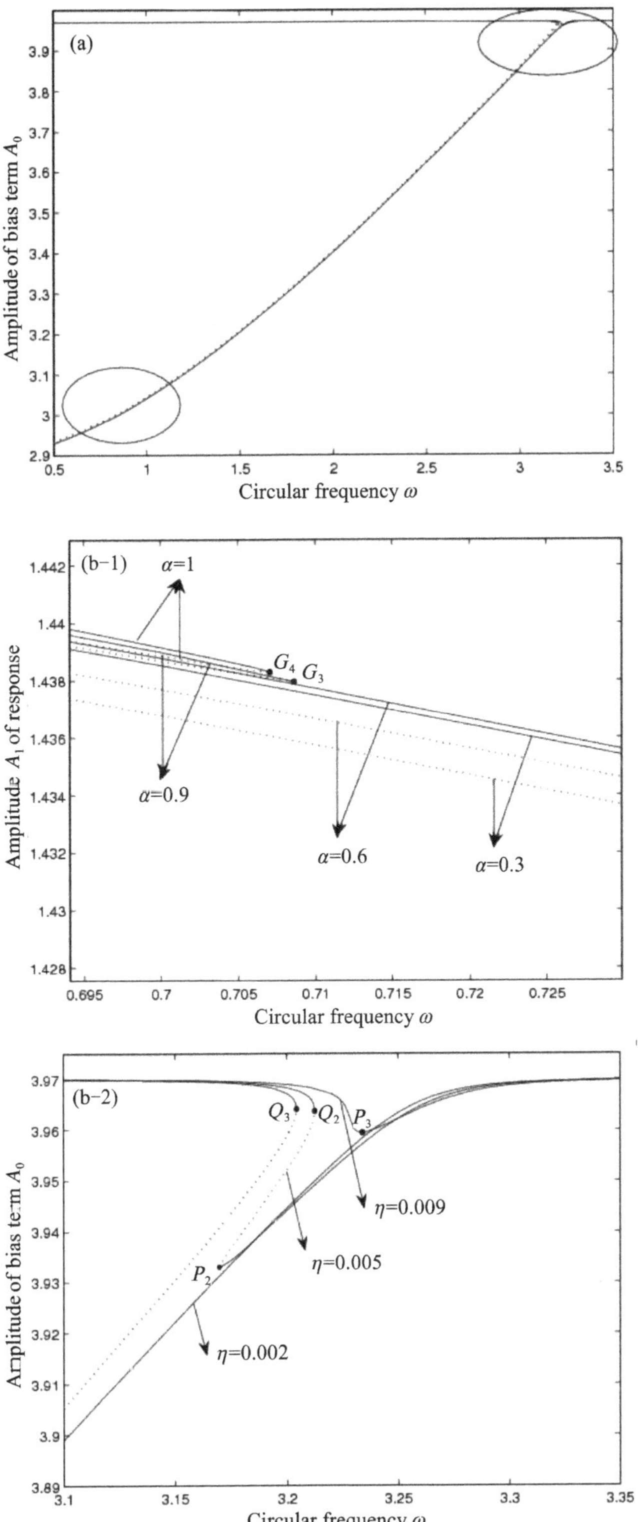

Figure 5.14 Upper region of response curves of bias terms A_0 for various material parameters: (a): panoramic view; (b-1)(b-2): local view

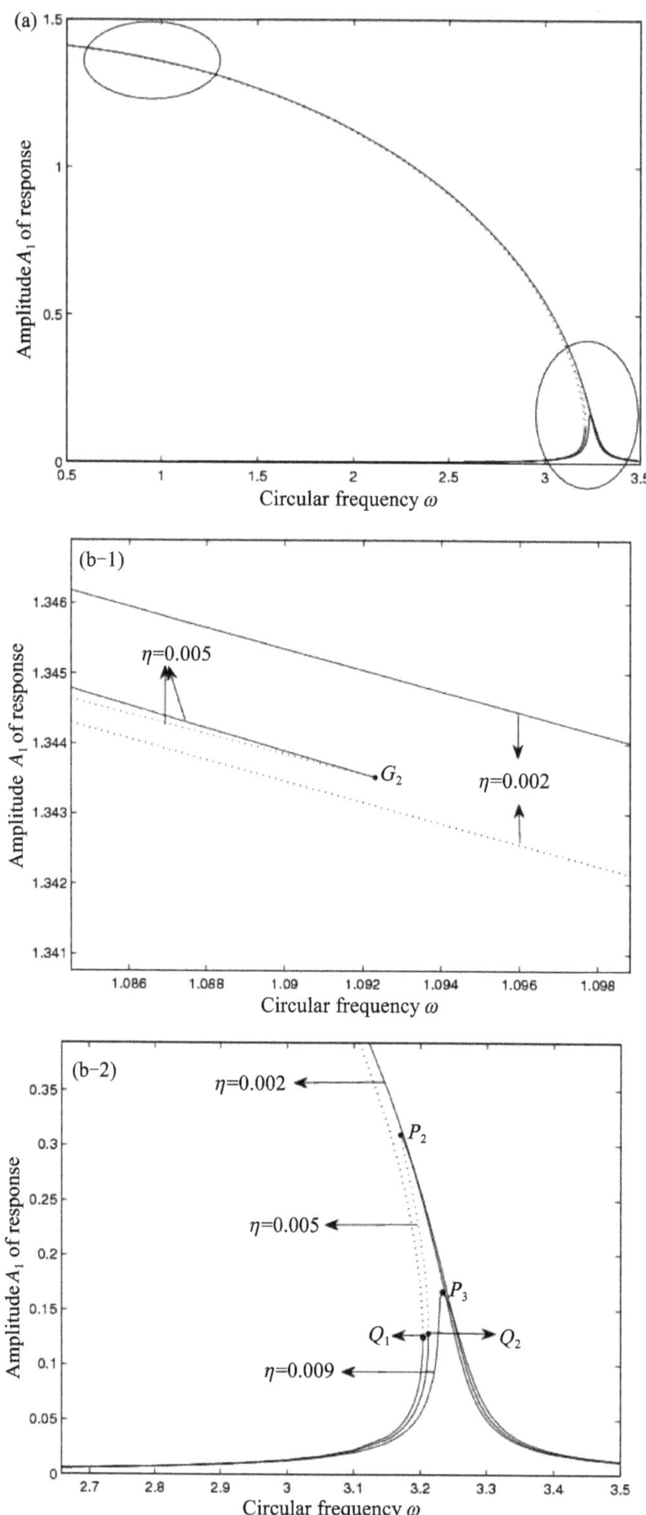

Figure 5.15 Upper region of response curves of amplitude A_1 for various material parameters:(a): Panoramic view; (b-1)(b-2): Local view

5.1.5 Conclusion

The nonlinear dynamic steady state responses depending on various control parameters are considered analytically without numerical integration. For certain values of the control parameters, coexisting of stable and unstable solutions give rise to hysteresis and jump phenomena. It is observed that the hysteresis domain shrinks according to the increases of the fractional order and the material modulus ratio. Additionally, the jump and hysteresis may be eliminated when the fractional order or the material modulus ratio exceeds some critical values and the mechanical systems can be stabilized by proper controlling the viscoelastic parameters.

5.2 Dynamic behavior of a viscoelastic plane truss system

5.2.1 Introduction

Truss like systems are very attractive in architecture or structural engineering, such as bridges, offshore platform, transmission towers, stage design and displays and even in the structures in space. The wide applicability of truss structures is mainly because that they combine high strength and stiffness with light weight. Although the system is simple, it is known to be extremely strong due to its capacity for dissipating forces throughout its structural members. Therefore, dynamic analysis for truss systems becomes very significant in system design and implement in practice.

Much research has been done to investigate the dynamic behavior of such systems in two and three dimensions [Cichon & Corradi, 1981; Papadrakakis, 1983; Kondoh & Atluri, 1985; Hill et al., 1989; De-Freitas & Ribeiro, 1992; Smith, E.M., 1994; Smith, M.E., 1994; Blandford, 1997]. Yang and Wu [Yang, Y.B. & Wu, 1997] investigated a symmetric two-member truss subjected to a vertical tip load. They found that the chaotic responses of the system can be identified either qualitatively by their sensitivity to initial conditions or the occurrence of strange attracters, or quantitatively by the existence of positive Lyapunov exponents or non-integer fractal dimensions. Thai and Kim [Thai & Kim, 2011] developed a numerical procedure through combining Newmark method with Newton-Raphson procedure and analyzed of nonlinear inelastic time-history truss system, for which both the geometric and material nonlinearities are considered.

The dynamic snap-through is an instability phenomenon that usually occurs in nonlinear analysis of some structures, such as the arch and truss systems. This instability behavior in general involves a sudden transition of the structural response from one equilibrium state to another non-adjoining equilibrium one when certain system parameter passes through a critical value. The sudden jump in deflection may indicate the snap-through of the structures. The snap-through behavior of the two-member truss under different dynamic loads was examined by Kassimali et al. [Kassimali & Bidhendi, 1988]. By studying a two-bar truss system, chaotic behavior was reported from the viewpoint of structural mechanics through numerical analysis of dynamic snap-through and homoclinic bifurcation [Ario, 2004]. Nevertheless, the nonlinear dynamics of a viscoelastic truss system with fractional order damping force has rarely been investigated.

In this section, based on the fractional Kelvin-Voigt model, the equations of a two-bar plane truss system are established and analyzed. The plane truss structure to be discussed in this research is assumed to be a general scheme and only a concentrated load at the nodes is considered. A fractional two-degree-of-freedom oscillator is obtained using one-term Galerkin truncation. The residue harmonic balance method along with polynomial homotopy continuation technique is used for all possible solutions of resulting nonlinear harmonic equations.

5.2.2 The Equations of Motion for Viscoelastic Plane Truss System

Consider a plane truss consisting of two pin-jointed straight bars sustaining plane loading shown in Figure 5.16. Two members of the truss are made of viscoelastic material whose constitutive behavior is described by the fractional Kelvin-Voigt model [Rossikhin, Y. A. & Shitikova, M. V., 2010]. Two members of the truss shown in Figure 5.16 have the cross section A, the density ρ, and length L_1 and L_2, respectively. Moreover, a span of H along X-direction and a vertical distance of V between Q and the Y-axis are assumed. The truss is subjected to a sinusoidal load of magnitude $F \sin \omega t$ at the tip with the angles of θ with respect to X-axis. However, only displacements within the XY plane are allowed. For convenience, only the nonlinearity resulting from the change in geometry of the truss is considered, while buckling of members in the Euler form will be completely excluded in this study.

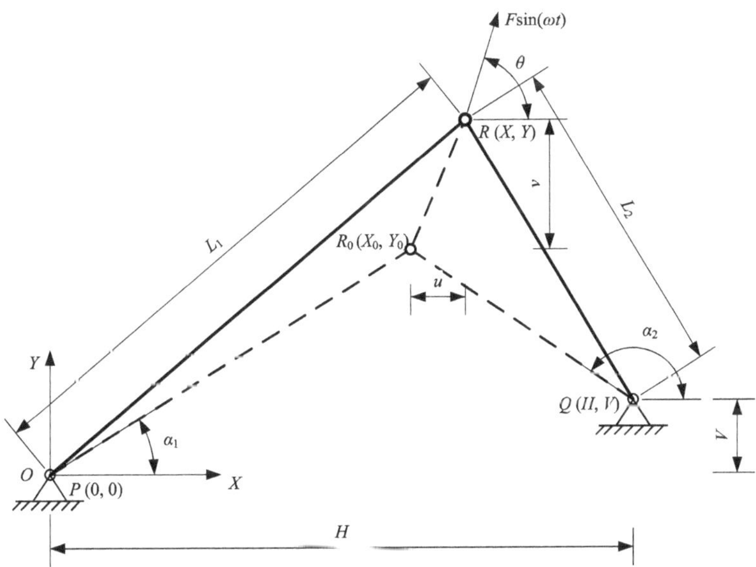

Figure 5.16 Schematic of viscoelastic two member truss under harmonic excitation

Original position of the node C is (X_0, Y_0), displacements in the axes are u and v. According to the definition of Green-Lagrange strain [Reddy, 2004b]:

$$\varepsilon = \frac{\partial u}{\partial X} + \frac{1}{2}\left[\left(\frac{\partial u}{\partial X}\right)^2 + \left(\frac{\partial v}{\partial X}\right)^2\right],\tag{5.19}$$

From [Leu & Yang, 1990; Yang, Y.B. & Wu, 1997], the nonlinear strain is

$$\varepsilon = e + \frac{1}{2}(e^2 + \theta^2) = \frac{\Delta u}{L_1} + \frac{1}{2}\left[\left(\frac{\Delta u}{L_1}\right)^2 + \left(\frac{\Delta v}{L_2}\right)^2\right],\tag{5.20}$$

where Δu and Δv are deformations along the axial and transverse direction. Deformation gradient tensor [Lai et al., 2010] is defined as $F = \nabla_0 u + I$, in which

$$[F] = \begin{bmatrix} \dfrac{\partial x}{\partial X} & \dfrac{\partial x}{\partial Y} \\[2mm] \dfrac{\partial y}{\partial X} & \dfrac{\partial y}{\partial Y} \end{bmatrix} = \begin{bmatrix} \dfrac{\Delta u}{L_1} & 0 \\[2mm] \dfrac{\Delta v}{L_2} & 0 \end{bmatrix} + \begin{bmatrix} 1 & 0 \\ 0 & 1 \end{bmatrix}.\tag{5.21}$$

The first Piola-Kirchhoff stress tensor gives the current force per unit undeformed area, while the second Piola-Kirchhoff stress tensor, used in the total Lagrangian formulation of geometrically nonlinear analysis, gives the transformed current force per unit undeformed area. Similar to $dX = F^{-1}dx$, the force df on the deformed element area da is transformed to the force dF on the undeformed element area dA [Reddy, 2004a], expressed as

$$dF = F^{-1}df \Rightarrow df = FdF.$$

And then,

$$\begin{Bmatrix} f_x \\ f_y \end{Bmatrix} = \begin{bmatrix} 1 + \dfrac{\Delta u}{L_1} & 0 \\[3mm] \dfrac{\Delta v}{L_2} & 1 \end{bmatrix}\begin{Bmatrix} F_X \\ 0 \end{Bmatrix} = \begin{Bmatrix} F_X\left(1 + \dfrac{\Delta u}{L_1}\right) \\[3mm] F_X\dfrac{\Delta v}{L_2} \end{Bmatrix}.\tag{5.22}$$

Here,

$$F_X = A(E_0[\varepsilon] + E_1 D_t^\alpha[\varepsilon]) = AE_0(1 + \eta D_t^\alpha)[\varepsilon],$$

where E_0 and E_1 refer to the elastic and viscoelastic parameter, respectively, and $\eta = E_1/E_0$ denotes the material modulus ratio. Therefore, in the global coordinate system, the internal force R can be obtained as follows.

$$\begin{pmatrix} R_X \\ R_Y \end{pmatrix} = \begin{pmatrix} R_X \\ R_Y \end{pmatrix}_1 + \begin{pmatrix} R_X \\ R_Y \end{pmatrix}_2, \tag{5.23}$$

where for the member I:

$$\begin{pmatrix} R_X \\ R_Y \end{pmatrix}_1 = \frac{1}{L_1} \begin{pmatrix} X \\ Y \end{pmatrix} AE_0(1+\eta D_t^\lambda)(\frac{X^2+Y^2-L_1^2}{2L_1^2}),$$

for the member II:

$$\begin{pmatrix} R_X \\ R_Y \end{pmatrix}_2 = \frac{1}{L_2} \begin{pmatrix} X-H \\ Y-V \end{pmatrix} AE_0(1+\eta D_t^\lambda)\left[\frac{(X-H)^2+(Y-V)^2-L_2^2}{2L_2^2}\right].$$

By the concept of lumped mass for the two members meeting at node C, equations of motion can be written as

$$\begin{cases} \dfrac{1}{2}\rho AL_1\ddot{X} + \dfrac{1}{2}\rho AL_2\ddot{X} + R_X = F_X\sin(\omega t), \\ \dfrac{1}{2}\rho AL_1\ddot{Y} + \dfrac{1}{2}\rho AL_2\ddot{Y} + R_Y = F_Y\sin(\omega t), \end{cases} \tag{5.24}$$

where $F_X = F\cos\theta$, $F_Y = F\sin\theta$. Then, we have

$$\begin{cases} \ddot{X} + \dfrac{2}{m_1+m_2}R_X = \dfrac{2F_X}{m_1+m_2}\sin(\omega t), \\ \ddot{Y} + \dfrac{2}{m_1+m_2}R_Y = \dfrac{2F_Y}{m_1+m_2}\sin(\omega t), \end{cases} \quad \text{with} \quad \begin{cases} m_1 = \rho AL_1, \\ m_2 = \rho AL_2. \end{cases} \tag{5.25}$$

Then, substituting R_X and R_Y in above equation, leads to

$$\rho A(L_1+L_2)X'' + AE_0\{\frac{1}{L_1^3}X(X^2+Y^2-L_1^2) + \frac{1}{L_2^3}(X-H)[(X-H)^2+(Y-V)^2-L_2^2]$$

$$+ \eta\frac{1}{L_1^3}XD_t^\lambda(X^2+Y^2) + \eta\frac{1}{L_2^3}(X-H)D_t^\lambda[(X-H)^2+(Y-V)^2]\} = 2\Gamma_X\cos(\omega t),$$

$$\rho A(L_1+L_2)Y'' + AE_0\{\frac{1}{L_1^3}Y(X^2+Y^2-L_1^2) + \frac{1}{L_2^3}(Y-V)[(X-H)^2+(Y-V)^2-L_2^2]$$

$$+ \eta\frac{1}{L_1^3}YD_t^\lambda(X^2+Y^2) + \eta\frac{1}{L_2^3}(Y-V)D_t^\lambda[(X-H)^2+(Y-V)^2]\} = 2F_Y\sin(\omega t),$$

$$\tag{5.26}$$

Let

$$u=\frac{X}{L_1},v=\frac{Y}{L_2},\frac{H}{L_2}=\overline{h}\varsigma,\frac{V}{L_2}=\overline{v}\varsigma,\frac{L_1}{L_2}=\varsigma,\zeta=\sqrt{\frac{E_0}{\rho}}\frac{t}{L_1+L_2}\overset{\triangle}{=}\vartheta t,$$

$$f_u=\frac{2F_X(L_1+L_2)}{AE_0L_1},f_v=\frac{2F_Y(L_1+L_2)}{AE_0L_2}.$$

Then, we obtain

$$\ddot{u}+(1+\varsigma^{-1})u(u^2+v^2\varsigma^{-2}-1)+(1+\varsigma)(u-\overline{h})[(u-\overline{h})^2\varsigma^2+(v-\overline{v}\varsigma)^2-1]$$
$$+\eta\vartheta^\lambda\{(1+\beta^{-1})uD_\tau^\lambda(u^2+v^2\varsigma^{-2})+(1+\varsigma)(u-\overline{h})D_\tau^\lambda[(u-\overline{h})^2\varsigma^2+(v-\overline{v}\varsigma)^2]\}$$
$$=f_u\sin(\omega t),$$
$$\ddot{v}+(1+\varsigma^{-1})v(u^2+v^2\varsigma^{-2}-1)+(1+\varsigma)(v-\overline{v}\varsigma)[(u-\overline{h})^2\varsigma^2+(v-\overline{v}\varsigma)^2-1]$$
$$+\eta\vartheta^\lambda\{(1+\varsigma^{-1})vD_\tau^\lambda(u^2+v^2\varsigma^{-2})+(1+\varsigma)(v-\overline{v}\varsigma)D_\tau^\lambda[(u-\overline{h})^2\varsigma^2+(v-\overline{v}\varsigma)^2]\}$$
$$=f_v\sin(\omega t),$$

$$(5.27)$$

where the over dot denotes the differentiation with respect to ζ. It is easy to find that

$$f_u=f\cos(\theta),f_v=f\sin(\theta),\tag{5.28}$$

after we introduce a variable transformation $f=\dfrac{2|F|(L_1+L_2)}{AE_0L_2}$, where $|F|$ is the magnitude of the external force $F(X,Y)$

5.2.3 Solution Procedure for Steady State Responses

The harmonic balance (HB) method is a commonly used method for time-periodic solutions for nonlinear dynamical systems. The first formal presentation of the harmonic balance method is usually credited to Kryloff and Bogoliuboff [Kryloff & Bogoliuboff, 1947]. Recently, it has been extended to fractional oscillators [Leung, Guo & Yang, 2012; Leung et al., 2013] in dynamic analysis, for which both the stable and unstable steady state solutions of the nonlinear fractional can be easily sought. However, when many harmonics are included in the classical HB method, the solution procedure becomes analytically cumbersome. The classical harmonic balance method is modified to treat the residues systematically to form the residue harmonic balance method. In this study, the residue harmonic balance method is then combined with the polynomial homotopy continuation [Sommese & Wampler, 2005] to solve the resulting nonlinear algebraic equations.

For convenience, we introduce two functions

$$\Phi_1 \triangleq \ddot{u} + (1+\varsigma^{-1})u(u^2 + v^2\varsigma^{-2} - 1) + (1+\varsigma)(u-\overline{h})[(u-\overline{h})^2\varsigma^2 + (v-\overline{v}\varsigma)^2 - 1]$$
$$+ \eta\vartheta^\lambda\{(1+\varsigma^{-1})uD_\tau^\lambda(u^2+v^2\varsigma^{-2}) + (1+\varsigma)(u-\overline{h})D_\tau^\lambda[(u-\overline{h})^2\varsigma^2 + (v-\overline{v}\varsigma)^2]\}$$
$$- f_u\sin(\omega t),$$

$$\Phi_2 \triangleq \ddot{v} + (1+\varsigma^{-1})v(u^2 + v^2\varsigma^{-2} - 1) + (1+\varsigma)(v-\overline{v}\varsigma)[(u-\overline{h})^2\varsigma^2 + (v-\overline{v}\varsigma)^2 - 1]$$
$$+ \eta\vartheta^\lambda\{(1+\varsigma^{-1})vD_\tau^\lambda(u^2+v^2\varsigma^{-2}) + (1+\varsigma)(v-\overline{v}\varsigma)D_\tau^\lambda[(u-\overline{h})^2\varsigma^2 + (v-\overline{v}\varsigma)^2]\}$$
$$- f_v\sin(\omega t).$$

$$(5.29)$$

For analysis of the primary resonance response, the solution of Eq. (5.27) is assumed to be

$$\begin{cases} u = a_0 + a_1\cos(\omega t) + b_1\sin(\omega t), \\ v = c_0 + c_1\cos(\omega t) + d_1\sin(\omega t). \end{cases} \tag{5.30}$$

Substituting Eq. (5.30) into Eq. (5.27), and equating the coefficients of like harmonic to zero yields a set of nonlinear algebraic equations with respect to Fourier coefficients in (5.30)

$$\begin{cases} R_i^0(a_0,a_1,b_1,c_0,c_1,d_1) = \dfrac{2}{\pi}\displaystyle\int_0^{\pi/\omega}\Phi_i\,d\tau = 0, i = 1,2, \\[2ex] R_i^c(a_0,a_1,b_1,c_0,c_1,d_1) = \dfrac{2}{\pi}\displaystyle\int_0^{\pi/\omega}\Phi_i\cos\omega t\,dt = 0, i = 1,2, \\[2ex] R_i^s(a_0,a_1,b_1,c_0,c_1,d_1) = \dfrac{2}{\pi}\displaystyle\int_0^{\pi/\omega}\Phi_i\sin\omega t\,dt = 0, i = 1,2. \end{cases} \tag{5.31}$$

5.2.4 Results and Discussions

In this section, the effects of the fractional order and material modulus ratio on the steady state response are studied by the aforementioned method using three harmonic terms in Eq. (5.30) In order for that, the amplitude response curves are illustratively presented. In these figures, the solid lines denote the stable solutions, while the dashed lines stand for unstable ones. The stability and instability of the responses are detected by using Runge-Kutta numerical integration where the initial conditions are supplied by the results of the harmonic balance approximation.

The material and structure parameters of the members of the viscoelastic truss system considered are given by

$$E_0 = 21\text{MPa}, \rho = 7.5 \times 10^3 \text{kg/m}^3,$$
$$L_1 = L_2 = 2\text{m}, H = 2.8\text{m}, V = 0.4\text{m}, A = 0.04\text{m}^2. \tag{5.32}$$

Moreover, the slant angle of the external force with respect to X-axis is fixed at $\theta = \pi/3$ in this study.

As usual, the steady state amplitudes of bias term and response are denoted, respectively, as follows:

$$A_0 = |a_0|, \, A_1 = \sqrt{a_1^2 + b_1^2}, \overline{A}_0 = |c_0|, \, \overline{A}_1 = \sqrt{c_1^2 + c_1^2}. \tag{5.33}$$

5.2.4.1 Effects of fractional order on amplitude-response curves

In this subsection, the effects of fractional order on the steady state responses are studied and illustratively presented by using amplitude-response curves in Figures 5.17–5.23, wherein, three case for $\alpha = 0.5, 0.8, 1$ are respectively considered.

In Figure 5.17, the relationship of amplitudes A_0 and $\overline{A}_0$ of bias term versus forcing amplitude f are ploted for $\alpha = 0.5$. It is easy to find that two response curves coexist on the forcing amplitude interval $f \in [0,2]$ For simplicity, we refer them as I and II respectively in the following context. These two curves emerge into each other at the intersection point R. Due to the presence of multiple solutions, saddle node bifurcations P,Q and P' exist on the response curves I and II , respectively. However, because the amplitudes in the response curve II are always equal to 0.7 and 0.1 for $u(t)$ and $v(t)$, respectively, ones can not obtain a visual representaion for bifurcating structure of the response curve II . Nevertheless, jumps can be observed between these two different rsponse curves, and then the presence of snap-through phenomenon is indicated.

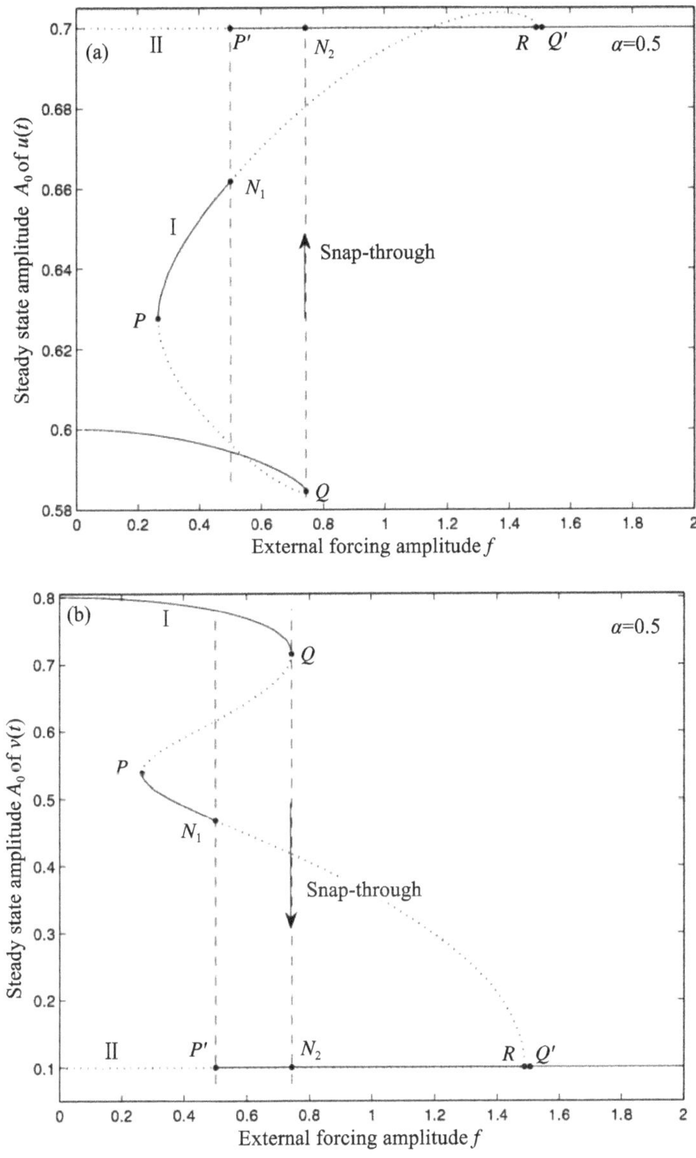

Figure 5.17 Steady state amplitude of the bias term of: (a) $u(t)$; (b) $v(t)$, for

$$\alpha = 0.5,\ \eta = 0.1,\ \omega = 1.2.$$

For detailed knowledge of snap-through phenomenon, the response amplitudes are ploted in Figure 5.18 as the function of forcing amplitude. It is worth stressing that, although the amplitude response curves in Figure 5.18 (a) and (b) respectively for $u(t)$ and $v(t)$ are very different in form, the bifurcating structures are consistent in nature. Hence, the examinations are focused on the response amplitude of $v(t)$ in Figure 5.18 (b).

From Figure 5.18 (b), it can be seen that only the solutions of the upper branch in the response curve II and the solutions between R and Q', and the solutions of lower brach as well as part of the upper branch between P and N_1 in the response I are stable. Therefore, the vibration configuration can change from the response curve I to response curve II at the saddle node Q by the dynamic snap-through as the forcing amplitude f increases. Specifically, as the forcing amplitude f increases, it reaches the saddle node Q on the response curve I . Increasing the forcing amplitde further, the vibration amplitude will jump to the point N_2 on the response curve II by the snap-through and then follow the upper branch of the curve II . Conversely, as the forcing amplitude decreases, the vibration amplitude reaches the saddle node P' on the response curve II , further decrease of the forcing amplitue leads to a jump down phenomenon from the response curve II at the saddle node P' to the response curve I at the point N_1, and then to the lower branch of the response curve I through a jump down at the saddle node P, which is usually called by "snap-back" phenonmenon. Howere, it should be mentioned that another way may be realised, i.e., jumping down directly from the response curve II at P' to the lower branch of the response curve I .

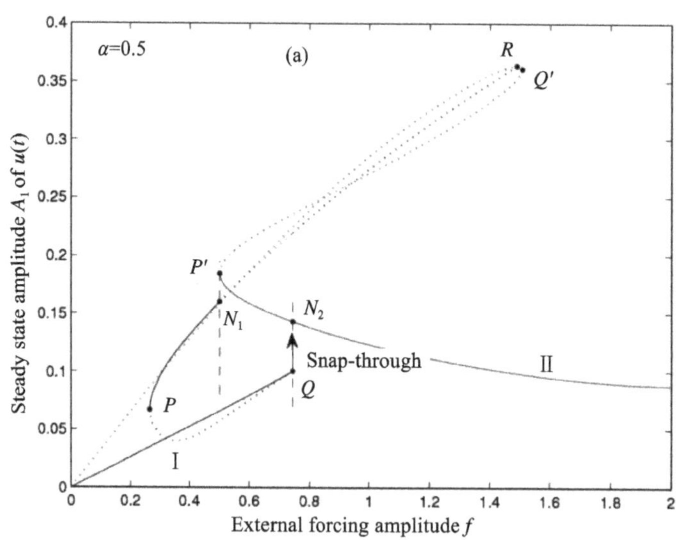

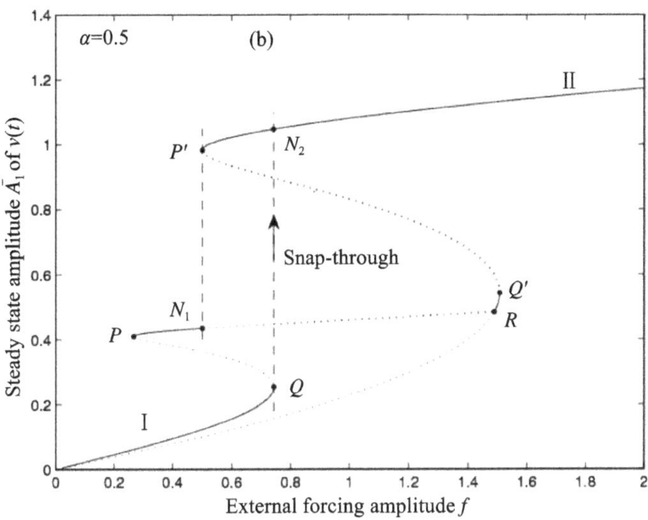

Figure 5.18 Steady state amplitude of the response (a): $u(t)$ **; (b):** $v(t)$**, for** $\alpha = 0.5$**,** $\eta = 0.1$**,** $\omega = 1.2$**.**

It should also be mentioned that as the forcing amplitude increases and reaches the intersection point R, the responses on the lower and upper baranches are all stable. The initia conditions determine how the system motion performs. Then, at the saddle node Q', a jump up occurs. It has been clearly illustrated in Figure 5.18 for response amplitudes of the system.

For the case of $\alpha = 0.8$, the amplitudes of the bias term and the system response are plotted in Figs. 5.19 and 5.20, respectively. For convenience in following discussions, the same notes of the bifurcation points and response curves are adopted as those in above discussions for the case of $\alpha = 0.5$.

Similarly, observations from Figure 5.19 and Figure 20 show that, two response curves I and II coexist and intersect at the point $R(f \approx 1.575703)$. The jumps and hysteresis phenomenon can also be found in response curves because of the presence of multiple solutions. Moreover, comparisons with the Figs. 5.17 and 5.18 indicate that, due to the increase of fractional order, hysteresis area of these two response curves are shrunk and delayed to a bigger level of forcing amplitude. However, different from Figure 5.17 and Figure 5.18 for $\alpha = 0.5$, two critical points $N_1(f \approx 1.223)$ and $N_2(\text{or } N_3)$ $(f \approx 1.5151)$ are found to determine an unstable region, on which the system exhibits more complex dynamics. Additionally, the snap-through phenomenon is also annihilated. However, due to the presence of complex and irregular motions, the system becomes undesirable.

To examine these complex motions, phase diagrams at some special values of forcing amplitude are ploted in Figure 5.21. Ones can find that with the increase of the forcing amplitude, the periodic-1 solutions lose the stability through a period-doubling bifurcation at N_1. The phase diagram for $f = 1.22$ and $f = 1.225$ is calculated and ploted in Figure 5.21 (a)-(b). And then, the system goes into a chaotic motion. The chaotic motion at $f = 1.312$ is ploted in Figure 5.21 (c). In addition, with the decrease of external force, the stable periodic motion (on the upper or lower brachch) loses stability and bifurcates to period-three motion, then enters chaotic motions. The period-three solution is given by Figure 5.21 (d) at $f = 1.51$.

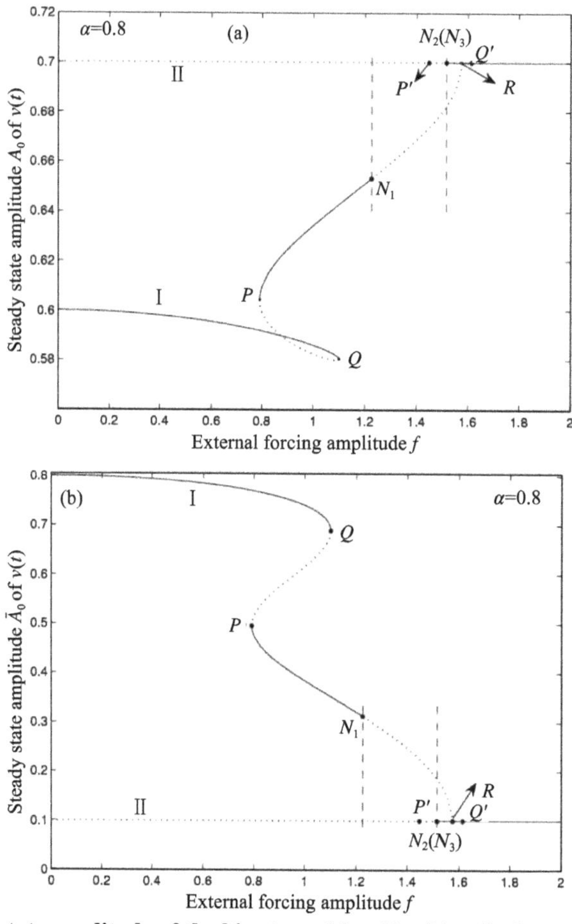

Figure 5.19 Steady state amplitude of the bias term (a): $u(t)$; (b): $v(t)$, for $\alpha = 0.8$, $\eta = 0.1$, $\omega = 1.2$.

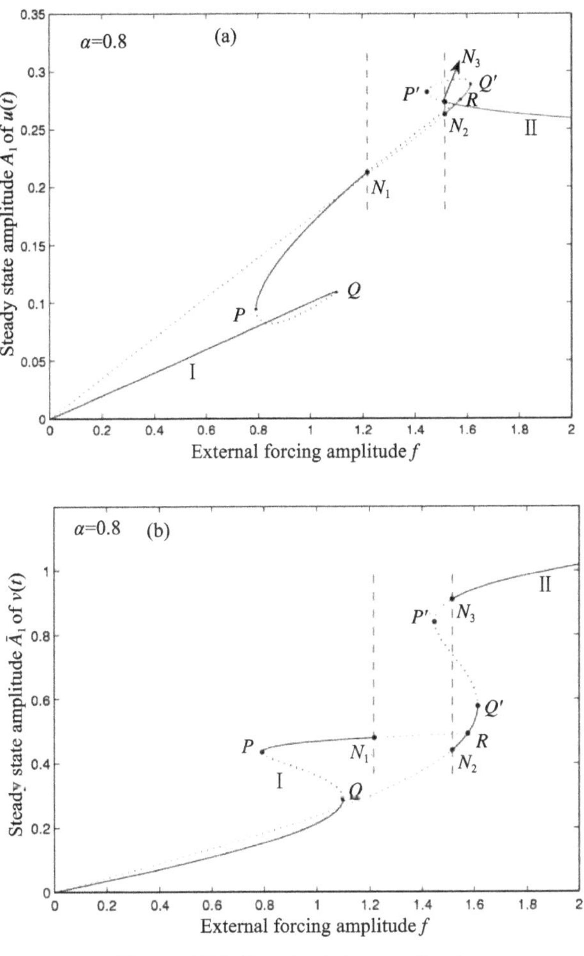

Figure 5.20 Steady state amplitude

(a): $u(t)$; **(b):** $v(t)$, for $\alpha = 0.8$, $\eta = 0.1$, $\omega = 1.2$

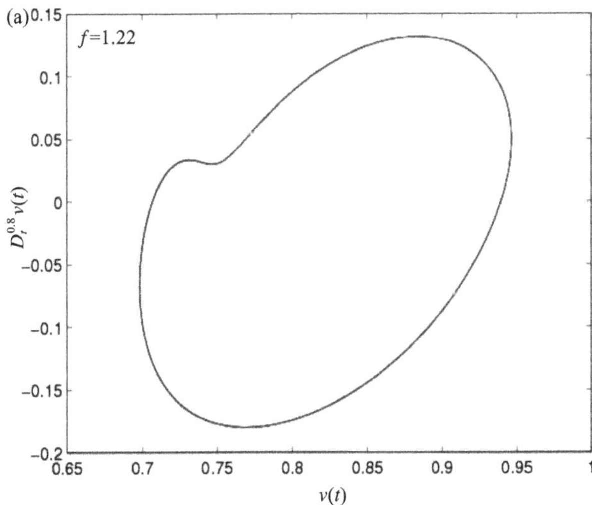

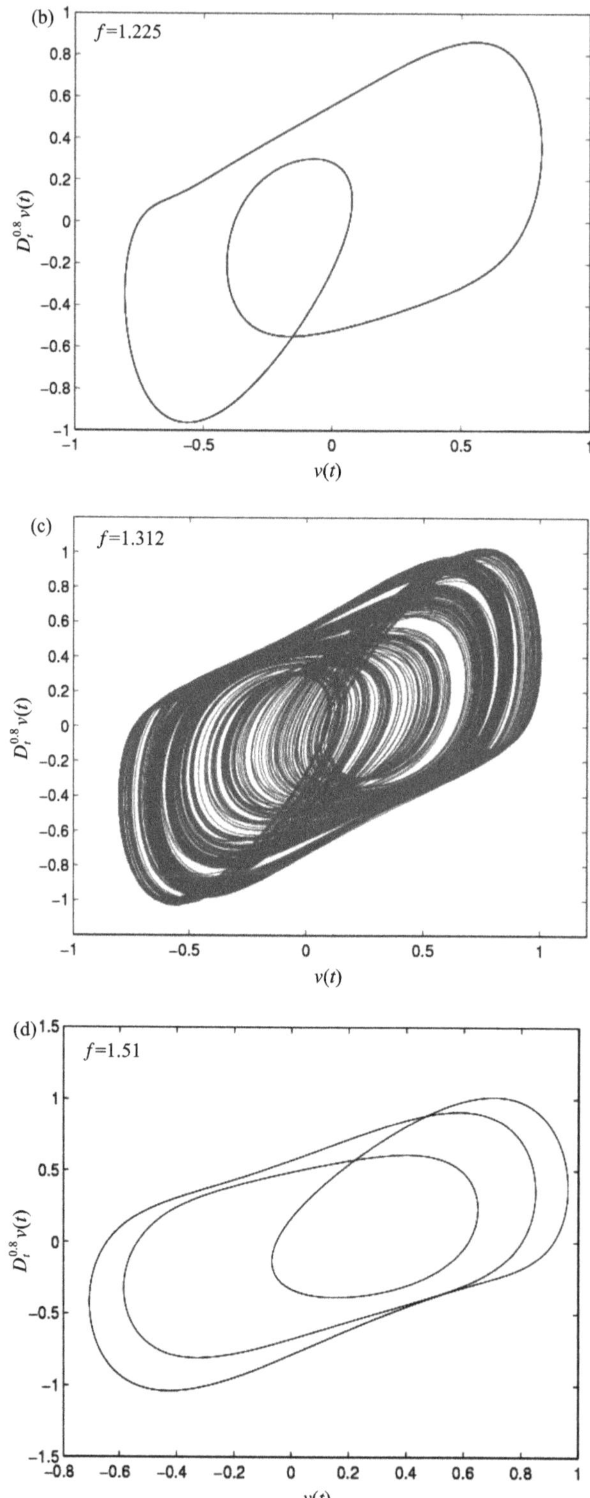

Figure 5.21 Phase portraits for $\alpha=0.8$, $\eta=0.1$, $\omega=1.2$

(a): Period-2 solution; (b): Chaotic motion; (c): Period-3 solution

The steady state amplitudes of bias terms and system responses for $\alpha = 1$ are pictured in Figure 5.22 and Figure 5.23, respectively. It is easy to find that the saddle nodes and the jump phenomenon on the response curve II are eliminated. Moreover, the hysteresis area associated with the response curve I is delayed and shrunk simultaneously. The snap-through phenomenon appears again at the saddle node $Q(f \approx 1.8653)$ and the intersection point $R(f \approx 1.83669)$, respectively.

Furthermore, from the Figs. 5.18, 5.20 and 5.23, ones may also find that the response amplitudes A_1 and $\overline{A}_1$ of $u(t)$ and $v(t)$ are suppressed with increasing of the fractional order. To summarize, the fractional order has given much benefit to stabilization of the system.

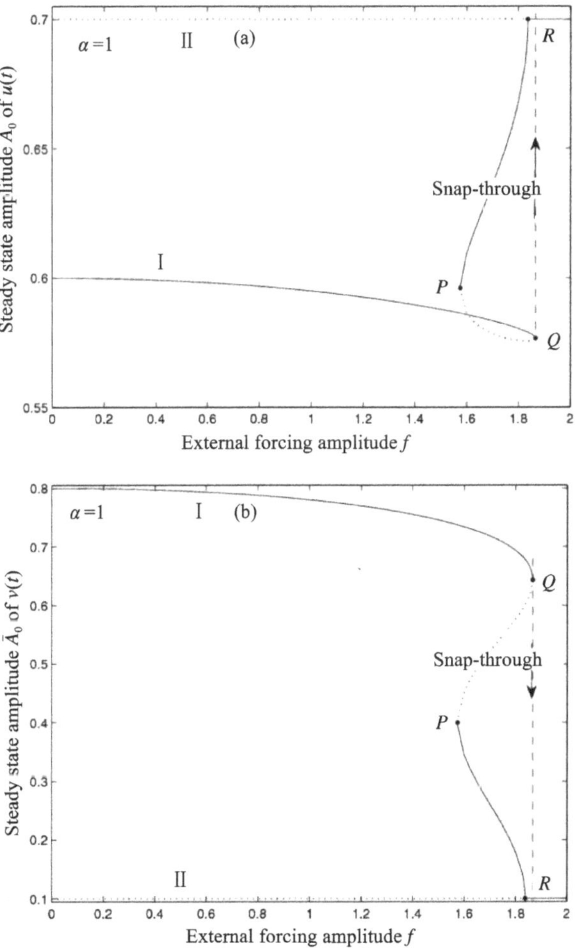

Figure 5.22 Steady state amplitude of the bias term

(a): $u(t)$; (b): $v(t)$, for $\alpha = 1, \eta = 0.1, \omega = 1.2$

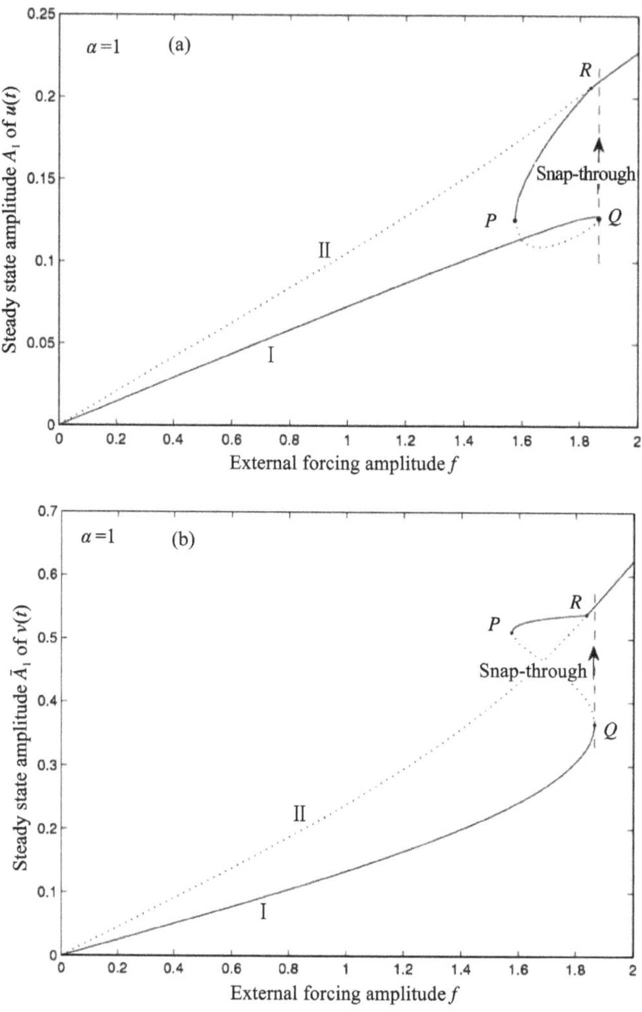

Figure 5.23 Steady state amplitude

(a) $u(t)$; (b) $v(t)$, for $\alpha=1$, $\eta=0.1$, $\omega=1.2$.

5.2.4.2 Effects of Material Modulus Ratio on Amplitude-Response Curves

In this subsection, the effects of material modulus ratio η on system performance are studied for $\eta=0.08,0.1,0.14$ by using amplitude response curves in Figure 5.24, 5.20 and 5.25, respectively.

The response curves in Figure 5.24 and Figure 5.20 are very similar in form to each other except that the locations of the bifurcation points on them have been delayed as the material modulus ratio η is increased. Similarly, because of the appearence of the complex motions in Figure 5.24, the dynamic snap-through does not happen in these two cases.

Further increase of η, for example in Figure 5.25 at $\eta = 0.14$, the saddle node and jumps on the response curve II are eliminated. A narrow band of excitation amplitude f : $f(N_1) = 1.4451 < f < f(N_2) = 1.5486$ is found, on which the complex motions occur. Therefore, the snap-through cannot be found as well. But any way, the system is not so desired due to the appearence of irregular motions. However, just as discussed in last subsection, the system will be stabilized as the material modulus ratio is further increased.

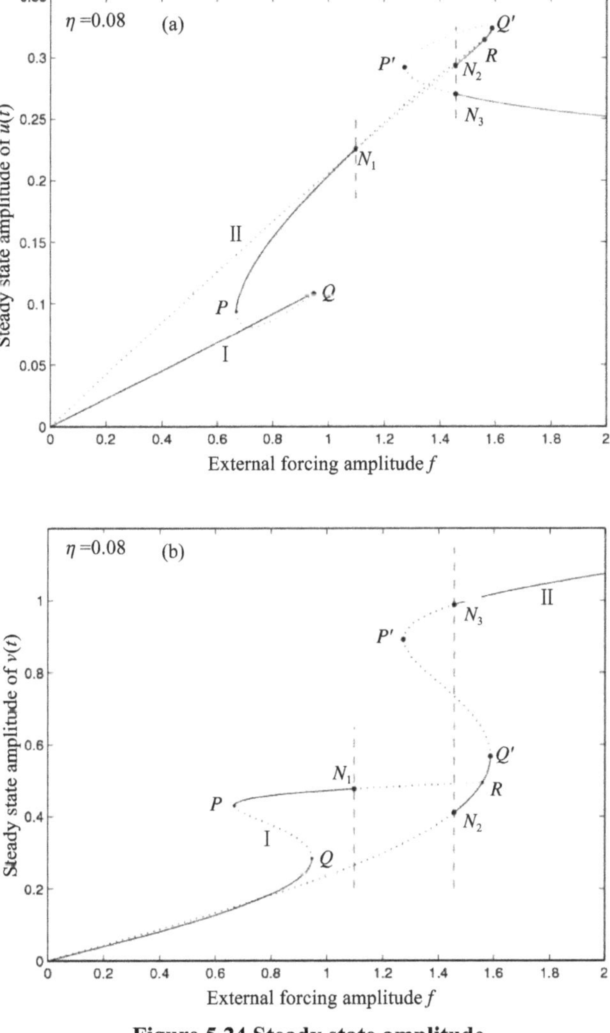

Figure 5.24 Steady state amplitude

(a): $u(t)$; (b): $v(t)$, for $\alpha = 0.8$, $\eta = 0.08$, $\omega = 1.2$

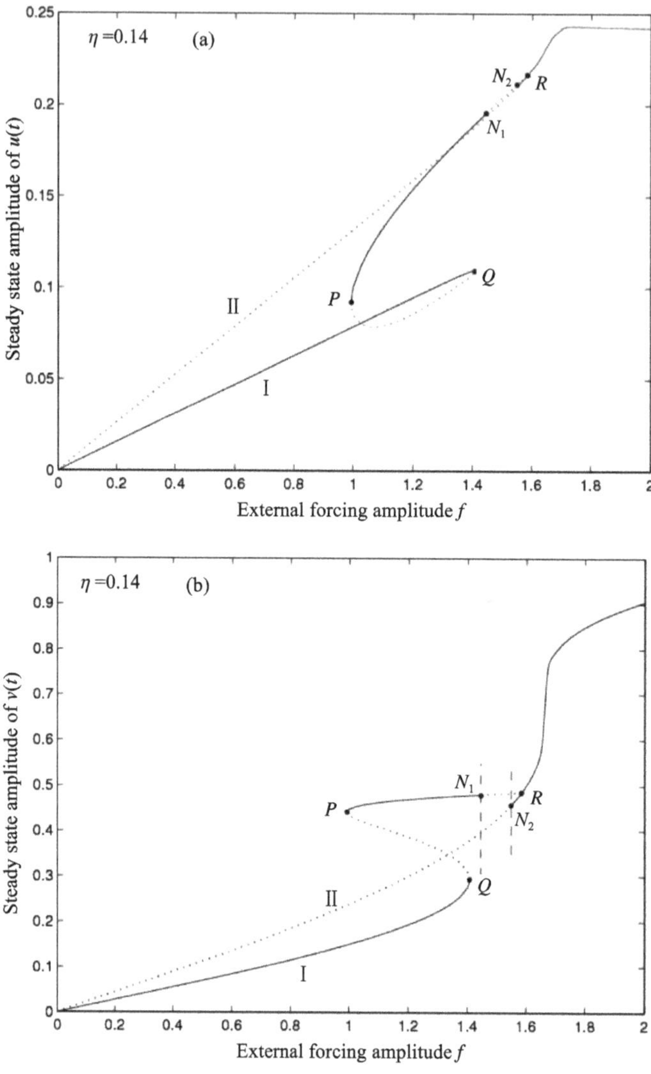

Figure 5.25 Steady state amplitude

(a): $u(t)$; (b): $v(t)$, for $\alpha = 0.8$, $\eta = 0.14$, $\omega = 1.2$

5.2.5 Conclusions

The dynamic behavior of a viscoelastic plane truss system is considered in this section. The constitutive behavior of the truss is characteristiced by the fractional Kelvin-Voigt model. The effects of fractional order and material modulus ratio are studies by using amplitude-response curves. For amplitude-response curves, fractional order (or material modulus ratio) has some critical values, beyond which the jump phenomena on the response curve II have been eliminated. In other words, both of them can give benefits to stabilizations of system.

When the external forcing amplitude goes over caritical value, the system exhibits the snap-through phenomena due to structural instability, and the motion is attracted to another equilibrium configuraction. However, it is found that the snap-through may disappear due to the presence of more complex motions.

APPENDICES

Appendix A

Fractional order derivative

The fractional-order derivative was defined by extending the concept of integer-order derivative. Three commonly used definitions for fractional-order derivative are Grunwald-Letnikov definition, Riemann-Liouville definition and Caputo definition [I. Podlubny, 1999].

When $0 < \alpha \notin N$, the α th order derivative of function $g(t)$ with respect to t and the terminal value zero is given as follows.

(i) Grunwald-Letnikov definition:

$$
{}^{GL}_{0}D_t^\alpha g(t) = \lim_{\substack{h \to 0 \\ nh=t}} h^{-\alpha} \sum_{r=0}^{n} (-1)^r \binom{\alpha}{r} g(t - rh)
\tag{1.3}
$$

where $\dbinom{\alpha}{r} = \dfrac{\Gamma(\alpha+1)}{r!\,\Gamma(\alpha-r+1)}$.

(ii) Riemann-Liouville definition:

$$
{}^{RL}_{0}D_t^\alpha g(t) = \frac{1}{\Gamma(m-\alpha)} \frac{\mathrm{d}^m}{\mathrm{d}t^m} \left(\int_0^t (t-\tau)^{m-\alpha-1} g(\tau)\,\mathrm{d}\tau \right)
\tag{1.4}
$$

where $m - 1 < \alpha < m$.

(iii) Caputo definition:

$$
{}^{C}_{0}D_t^\alpha g(t) = \frac{1}{\Gamma(m-\alpha)} \int_0^t (t-\tau)^{m-\alpha-1} g^{(m)}(\tau)\,\mathrm{d}\tau
\tag{1.5}
$$

where $m - 1 < \alpha < m$.

Let us turn our attention to the properties and conclusions of fractional-order differentiation.

(i) Linearity

$$D^\alpha(\lambda_1 g_1(t) + \lambda_2 g_2(t)) = \lambda_1 D^\alpha g_1(t) + \lambda_2 D^\alpha g_2(t) \tag{1.6}$$

where D^α denotes any mutation of the fractional differentiation mentioned above.

(ii) The Leibniz rule for fractional derivatives

$$^{RL}_0 D^\alpha_t(g_1(t)g_2(t)) = \sum_{k=0}^{\infty} \binom{\alpha}{k} g_1^{(k)}(t)\, ^{RL}_0 D^{\alpha-k}_t g_2(t) \tag{1.7}$$

(iii) Fractional derivative of a composite function

Suppose that $\varphi(t)$ is a composite function, $\varphi(t) = G(g(t))$, then

$$^{RL}_0 D^\alpha_t G(g(t)) = \frac{t^\alpha}{\Gamma(1-\alpha)}\varphi(t) + \sum_{k=1}^{\infty}\binom{\alpha}{k}\frac{k!\, t^{k-\alpha}}{\Gamma(k-\alpha+1)}\sum_{i=1}^{k} G^{(i)}(g(t))\sum\prod_{j=1}^{k}\frac{1}{\lambda_r!}\left(\frac{g^{(r)}(t)}{r!}\right)^{\lambda_r} \tag{1.8}$$

where the sum $\sum$ extends over all combinations of non-negative inter-values of

$\lambda_1, \lambda_2, \cdots, \lambda_k$ such that $\sum_{r=1}^{k} r\lambda_r = k$ and $\sum_{r=1}^{k}\lambda_r = m$

(iv) Interchange of the differentiation operators

$$^C_0 D^\alpha_t(^C_0 D^n_t g(t)) = \,^C_0 D^n_t(^C_0 D^\alpha_t g(t)) = \,^C_0 D^{\alpha+n}_t g(t),$$
$$g^{(s)}(0) = 0, s = k, k+1, \cdots, n; (n = 0,1,2,\cdots; k-1 < \alpha < k) \tag{1.9}$$

$$^{RL}_0 D^\alpha_t(^{RL}_0 D^n_t g(t)) = \,^{RL}_0 D^n_t(^{RL}_0 D^\alpha_t g(t)) = \,^{RL}_0 D^{\alpha+n}_t g(t),$$
$$g^{(s)}(0) = 0, s = 0,1,\cdots, n; (n = 0,1,2,\cdots; k-1 < \alpha < k) \tag{1.10}$$

In addition, the following important property has been proved [I. Podlubny, 1999].

If $g(t)$ is $(n-1)$-times continuously differentiable and $g(t)^{(n)}$ is integrable, then

$$^{GL}_0 D^\alpha_t g(t) = \,^{RL}_0 D^\alpha_t g(t) = \,^C_0 D^\alpha_t g(t) + \sum_{k=0}^{n-1}\frac{t^{k-\alpha}}{\Gamma(k-\alpha+1)} g^{(k)}(0^+) \tag{1.11}$$

where $n-1 < \alpha < n$.

Appendix B

The corresponding zeroth-, first- and second-order residue harmonic balance equations of Eq. (2.36) are as below:

Zeroth-order residue harmonic balance equation:

$$\omega_0^2 u_0'' + (h_1 + h_2 u_0^2)\omega_0^\alpha D_\tau^\alpha u_0 + \Omega_0^2 u_0 + k u_0^3 = R_1$$

where $R_1 = \dfrac{1}{4} a_0^3 (h_2 \omega_0^\alpha \cos\dfrac{\alpha\pi}{2} + k)\cos(3\tau) - \dfrac{1}{4} h_2 \omega_0^\alpha a_0^3 \sin\dfrac{\alpha\pi}{2}\sin(3\tau)$.

First-order residue harmonic balance equation:

$$\omega_0^2 u_1'' + 2\omega_0\omega_1 u_0'' + (h_1 + h_2 u_0^2)[\omega_0^\alpha D_\tau^\alpha u_1 + \alpha\omega_0^{\alpha-1}\omega_1 D_\tau^\alpha u_0] + 2h_2 u_0 u_1 \omega_0^\alpha D_\tau^\alpha u_0 + \Omega_0^2 u_1 + 3k u_0^2 u_1$$

$$+ R_1 = R_2$$

where

$$R_2 = \frac{1}{4}[(2+3^\alpha)h_2\omega_0^\alpha a_0^2(a_{3,1}\cos\frac{\alpha\pi}{2} + b_{3,1}\sin\frac{\alpha\pi}{2}) + 3k a_0^2 a_{3,1}]\cos(5\tau)$$

$$+ \frac{1}{4}[(2+3^\alpha)h_2\omega_0^\alpha a_0^2(b_{3,1}\cos\frac{\alpha\pi}{2} - a_{3,1}\sin\frac{\alpha\pi}{2}) + 3k a_0^2 b_{3,1}]\sin(5\tau).$$

Second-order residue harmonic balance equation:

$$\omega_0^2 u_2'' + 2\omega_0\omega_1 u_1'' + (2\omega_0\omega_2 + \omega_1^2)u_0'' + (h_1 + h_2 u_0^2)\{\omega_0^\alpha D_\tau^\alpha u_2 + \alpha\omega_0^{\alpha-1}\omega_1 D_\tau^\alpha u_1$$

$$+[\alpha\omega_0^{\alpha-1}\omega_2 + \frac{\alpha(\alpha-1)}{2}\omega_0^{\alpha-2}\omega_1^2]D_\tau^\alpha u_0\} + 2h_2 u_0 u_1(\omega_0^\alpha D_\tau^\alpha u_1 + \alpha\omega_0^{\alpha-1}\omega_1 D_\tau^\alpha u_0)$$

$$+h_2(2u_0 u_2 + u_1^2)\omega_0^\alpha D_\tau^\alpha u_0 + h_0^2 u_2 + 3k u_0^2 u_2 + 3k u_0 u_1^2 + R_2 = R_3.$$

Appendix C

The corresponding harmonic balance equations (4.22) when $n=1$ in solution (4.18) as follows:

$$\Phi_1 = \omega_1^2 a_0 + h_2 g_1 n_1 a_0(b_1^2 + a_1^2) - \frac{1}{2}f_2 a_1 + k_1(a_0^2 + \frac{1}{2}a_1^2 + \frac{1}{2}b_1^2) + k_2(a_0^3 + \frac{3}{2}a_0 b_1^2$$

$$+\frac{3}{2}a_0 a_1^2) + k_3 c_0 + k_4(a_0^2 - 2a_0 c_0 + \frac{1}{2}a_1^2 + \frac{1}{2}b_1^2 + \frac{1}{2}c_1^2 + \frac{1}{2}d_1^2 - a_1 c_1 - b_1 d_1 + c_0^2)$$

$$+k_5(3a_0 c_0^2 - 3a_0^2 c_0 + 3a_1 c_0 c_1 + 3b_1 c_0 d_1 - 3a_0 a_1 c_1 - \frac{3}{2}a_1^2 c_0 + \frac{3}{2}a_0 b_1^2 + \frac{3}{2}a_0 a_1^2 - \frac{3}{2}c_0 c_1^2$$

$$+\frac{3}{2}a_0 c_1^2 + \frac{3}{2}a_0 d_1^2 - \frac{3}{2}c_0 b_1^2 - \frac{3}{2}c_0 d_1^2 + a_0^3 - c_0^3 - 3a_0 b_1 d_1)$$

$$\Phi_2 = -f_1 - f_2 a_0 + \omega_1^2 a_1 - \omega^2 a_1 + h_1 g_1(n_1 a_1 + m_1 b_1) + h_2 g_1(\frac{3}{4} n_1 a_1^3 + \frac{1}{4} m_1 b_1^3 +$$

$$\frac{3}{4} n_1 a_1 b_1^2 + \frac{1}{4} m_1 a_1^2 b_1 + m_1 a_0^2 b_1 + n_1 a_0^2 a_1) + h_3 g_1(-m_1 d_1 - n_1 c_1 + n_1 a_1 + m_1 b_1) +$$

$$k_1 a_0 a_1 + k_2(\frac{3}{4} a_1 b_1^2 + 3a_0^2 a_1 + \frac{3}{4} a_1^3) + k_3 c_1 + k_4(-2a_0 c_1 - 2a_1 c_0 + 2c_0 c_1 + 2a_0 a_1) +$$

$$k_5(6a_0 c_0 c_1 - 6a_0 a_1 c_0 + \frac{3}{4} a_1^3 - \frac{3}{4} c_1^3 - \frac{3}{4} b_1^2 c_1 - \frac{9}{4} a_1^2 c_1 - \frac{3}{4} c_1 d_1^2 + \frac{3}{4} a_1 b_1^2 + \frac{3}{4} a_1 d_1^2$$

$$+\frac{9}{4} a_1 c_1^2 + \frac{2}{3} b_1 c_1 d_1 + 3a_0^2 a_1 - 3a_0^2 c_1 + 3a_1 c_0^2 - 3c_0^2 c_1 - \frac{3}{2} a_1 b_1 d_1)$$

$$\Phi_3 = \omega_1^2 b_1 - \omega^2 b_1 + h_1 g_1(n_1 b_1 - m_1 a_1) + h_2 g_1(-m_1 a_0^2 a_1 - \frac{1}{4} m_1 a_1^3 + \frac{3}{4} n_1 b_1^3 - \frac{1}{4} m_1 a_1 b_1^2 + \frac{3}{4} n_1 a_1^2 b_1$$

$$+n_1 a_0^2 b_1) + h_3 g_1(m_1 c_1 - n_1 d_1 + n_1 b_1 - m_1 a_1) + k_1 a_0 b_1 + k_2(3a_0^2 b_1 + \frac{3}{4} a_1^2 b_1 + \frac{3}{4} b_1^3) + k_3 d_1 + k_4$$

$$(-a_0 d_1 - b_1 c_0 + 2c_0 d_1 + 2a_0 b_1) + k_5(\frac{3}{4} b_1 c_1^2 + a_0^2 b_1 + \frac{3}{4} a_1^2 b_1 + \frac{3}{2} a_1 c_1 d_1 + \frac{9}{4} b_1 d_1^2 + \frac{3}{4} b_1^3 - \frac{3}{4} d_1^3 -$$

$$\frac{9}{4} b_1^2 d_1 - a_0^2 d_1 + 3b_1 c_0^2 - 3c_0^2 d_1 - 6a_0 b_1 c_0 + 6a_0 d_1 c_0 - \frac{3}{4} a_1^2 d_1 - \frac{3}{2} a_1 b_1 c_1 - \frac{3}{4} c_1^2 d_1)$$

$$\Phi_4 = \omega_2^2 c_0 - \omega_2^2 a_0 + k_6(b_1 d_1 + \frac{1}{2} d_1^2 + \frac{1}{2} b_1^2 - 2a_0 c_0 + a_0^2 + c_0^2 + \frac{1}{2} a_1^2 + \frac{1}{2} c_1^2 - a_1 c_1) + k_7(c_0^3$$

$$-a_0^3 - 3a_0 c_0^2 + a_0^2 c_0 + \frac{3}{2} c_0 d_1^2 - \frac{3}{2} a_0 d_1^2 + \frac{3}{2} a_1^2 c_0 - \frac{3}{2} a_0 a_1^2 - \frac{3}{2} a_0 c_1^2 + 3a_0 a_1 c_1 - 3a_1 c_0 c_1 - \frac{3}{2} a_0 b_1^2$$

$$+\frac{3}{2} b_1^2 c_0 + 3a_0 b_1 d_1 - 3b_1 c_0 d_1 + \frac{3}{2} c_0 c_1^2)$$

$$\Phi_5 = -\omega^2 c_1 + \omega_2^2 c_1 - \omega_2^2 a_1 + h_4 g_1(m_1 d_1 + n_1 c_1 - m_1 b_1 - n_1 a_1) + k_6(2c_0 c_1 - 2a_1 c_0$$

$$+2a_0 a_1 - 2a_0 c_1) + k_7(6a_0 a_1 c_0 - 6a_0 c_0 c_1 + \frac{3}{4} c_1^3 - \frac{3}{4} a_1^3 + \frac{3}{2} a_1 b_1 d_1 - \frac{3}{2} b_1 c_1 d_1 - \frac{9}{4} a_1 c_1^2$$

$$+\frac{9}{4} a_1^2 c_1 + 3a_0^2 c_1 + 3c_0^2 c_1 - 3c_0^2 a_1 - 3a_0^2 a_1 - \frac{3}{4} a_1 d_1^2 + \frac{3}{4} c_1 d_1^2 + \frac{3}{4} b_1^2 c_1 - \frac{3}{4} a_1 b_1^2)$$

$$\Phi_6 = -\omega^2 d_1 + \omega_2^2 d_1 - \omega_2^2 b_1 + h_4 g_1(n_1 d_1 + m_1 a_1 - m_1 c_1 - n_1 b_1) + k_6(2a_0 b_1 - 2b_1 c_0$$

$$-2a_0 d_1 + 2c_0 d_1 + k_7(-6a_0 c_0 d_1 - \frac{9}{4} b_1 d_1^2 + 6a_0 b_1 c_0 - \frac{3}{4} b_1^3 + \frac{3}{2} a_1 b_1 c_1 - \frac{3}{2} a_1 c_1 d_1 +$$

$$\frac{9}{4} b_1^2 d_1 - \frac{3}{4} b_1 c_1 + 3c_0^2 d_1 - 3b_1 c_0^2 - 3a_0^2 b_1 + \frac{3}{4} a_1^2 d_1 + \frac{3}{4} d_1^3 - \frac{3}{4} a_1^2 b_1 + \frac{3}{4} c_1^2 d_1 + 3a_0^2 d_1)$$

where $m_1 = \cos(\dfrac{\lambda\pi}{2}), n_1 = \sin(\dfrac{\lambda\pi}{2})$ and $g_1 = \omega^\lambda$.

Appendix D: Polynomial homotopy continuation technique

Polynomial homotopy continuation technique

The polynomial continuation method aims at finding possible all solutions reliably rather than on finding one or a few solutions. It is useful on a wide variety of problems which exhibit a continuous dependence of the solutions on the continuation parameters. For convenience, we rewrite nonlinear Eq. (4.24) in a general form

$$\Phi(q) = [\Phi_1(q), \Phi_2(q), \cdots, \Phi_{4n+2}(q)]^T = 0 \tag{4.25}$$

where $q = (a_0, a_1, b_1, \cdots, a_n, b_n, c_0, c_1, d_1, \cdots, c_n, d_n)$ consists of the unknowns.

For the nonlinear algebraic Eq. (4.25), we consider the system of $4n+3$ equations in $4n+3$ unknowns after homogenization by introducing a parameter χ, the equation is homogenized to

$$\tilde{\Phi}: \begin{cases} \tilde{\Phi}_1(a_0, a_1, b_1, \cdots, c_n, d_n, \chi) = 0 \\ \quad\quad\vdots \\ \tilde{\Phi}_{4n+2}(a_0, a_1, b_1, \cdots, c_n, d_n, \chi) = 0 \\ l_1 a_0 + l_2 a_1 + l_3 b_1 + \cdots + l_{4n+1} c_n + l_{4n+2} d_n + l_{4n+3}\chi = 1 \end{cases} \tag{4.26}$$

whereare $l_i (i = 1, 2, \cdots, 4n+3)$ complex numbers. For the Eq. (4.25), we first construct a new equation by omitting some of the terms and adding new ones

$$\varphi(q) = [\varphi_1(q), \varphi_2(q), \cdots, \varphi_{4n+2}(q)]^T = 0 \tag{4.27}$$

that has exactly the same number of regular solutions as Eq. (4.25). Then homogenize φ into $\tilde{\varphi}$ as that have been done for Eq. (4.26). Then the homotopy continuation procedure is carried out to get the homotopy equation

The next step, we follow all the solution paths of the homotopy equation by introducing a homotopy parameter η,

$$\tilde{H}(q, \chi, \eta) = (1-\eta)\sigma\tilde{\varphi}(q, \chi) + \eta\tilde{\Phi}(q, \chi) \tag{4.28}$$

where is a set of coefficients and the paths stay in the $4n+3$ dimensions complex space for $\chi \neq 0$. Secondly, as the homotopy parameter η moves from 0 to 1, use numerical continuation to trace the paths that originate at the solutions of the start system towards the solutions of the target system. When $\chi = 1$, the solution of Eq. (4.26) reduces to that of Eq. (4.25). The properties of triviality ensure smoothness and accessibility. The solution paths can be followed from the initial points at $\eta = 0$ to the all solutions of Eq. (4.25). Wu [2005; 2006] provided some useful choices about the auxiliary homotopy function. In general, we modify the equations by omitting some of the terms from original system until we have a new start system of equations and the solutions to which may be easily guessed. We then deform the coefficients of the start system into the coefficients of the original system by a series of small increments to obtain our solutions.

References

Abdulaziz, O. et al. (2008) Solving systems of fractional differential equations by homotopy perturbation method, *Phys Lett A* **372:4**, 451-459.

Achar, B.N.N. et al. (2002) Response characteristics of a fractional oscillator, *Physica A* **309**, 275-288.

Achar, B.N.N. et al. (2004) Damping characteristics of a fractional oscillator, *Physica A* **339**, 311-319.

Achar, B.N.N. et al. (2001) Dynamics of the fractional oscillator, *Physica A* **297**, 361-367.

Adolfsson, K. & Enelund, M. (2003) Fractional derivative viscoelasticity at large deformations, *Nonlinear Dynam* **33**, 301-321.

Agrawal, S.K. et al. (2012) Synchronization of fractional order chaotic systems using active control method, *Chaos Soliton Fract* **45**, 737-752.

Ahmad, W. et al. (2001) Fractional-order Wien-bridge oscillator, *Electron Lett* **37**, 1110-1112.

Ahmad, W.M. & Sprott, J.C. (2003) Chaos in fractional-order autonomous nonlinear systems, *Chaos Soliton Fract* **16**, 339-351.

Ahmed, E. et al. (2006) On some Routh-Hurwitz conditions for fractional order differential equations and their applications in Lorenz, Rossler, Chua and Chen systems, *Phys Lett A* **358**, 1-4.

Alomari, A.K. et al. (2010) Homotopy analysis method for solving fractional Lorenz system, *Commun Nonlinear Sci* **15**, 1864-1872.

Amer, Y.A. & Bauomy, H.S. (2009) Vibration reduction in a 2DOF twin-tail system to parametric excitations, *Commun Nonlinear Sci* **14**, 560-573.

Arena, P. et al. (2000) *Nonlinear noninteger order circuits and systems—an introduction* (World Scientific, Singapore).

Arikoglu, A. & Ozkol, I. (2007) Solution of fractional differential equations by using differential transform method, *Chaos Soliton Fract* **34**, 1473-1481.

Arikoglu, A. & Ozkol, I. (2009) Solution of fractional integro-differential equations by using fractional differential transform method, *Chaos Soliton Fract* **40**, 521-529.

Ario, I. (2004) Homoclinic bifurcation and chaos attractor in elastic two-bar truss, Int J *Nonlin Mech* **39**, 605-617.

Atay, F.M. (1998) Van der Pol's oscillator under delayed feedback, *J Sound Vib* **218**, 333-339.

Attari, M. et al. (2010a) Analysis of a fractional order van der Pol-like oscillator via describing function method, *Nonlinear Dynam* **61**, 265-274.

Attari, M. et al. (2010b) Analysis of a fractional order Van der Pol-like oscillator via describing function method, *Nonlinear Dynam* **61**, 265-274.

Babouskos, N.G. & Katsikadelis, J.T. (2010) Nonlinear vibrations of viscoelastic plates of fractional derivative type: an AEM solution, *The Open Mechanics Journal* **4**, 8-20.

Baleanu, D., Diethelm, K., Scalas, E. & Trujillo, J.J. (2012) *Fractional calculus: models and numerical methods*, World Scientific.

Bagley, R.L. (1989) Power law and fractional calculus model of viscoelasticity, *AIAA Journal* **27**, 1412-1417.

Bagley, R.L. & Calico, R.A. (1991) Fractional order state-equations for the control of viscoelastically damped structures, *J Guid Control Dynam* **14**, 304-311.

Bagley, R.L. & Torvik, P.J. (1983a) Fractional calculus-A different approach to the analysis of viscoelastically damped structures, *AIAA Journal* **21**, 741-748.

Bagley, R.L. & Torvik, P.J. (1983b) A theoretical basis for the application of fractional calculus to viscoelasticity, *Journal of Rheology* **27**, 201-210.

Bagley, R.L. & Torvik, P.J. (1985) Fractional calculus in the transient analysis of viscoelastically damped structures, *AIAA Journal* **23**, 918-925.

Bahraini, S.M.S. et al. (2013) Large deflection of viscoelastic beams using fractional derivative model, Journal of *Mechanical Science and Technology* **27**, 1063-1070.

Baleanu, D. et al. (2012a) *Fractional calculus models and numerical methods, series on complexity, nonlinearity and chaos* (World Scientific, Singapore).

Baleanu, D. et al. (2012b). Fractional calculus: models and numerical methods, *Proc. Complexity, Nonlinearity and Chaos*, ed. Luo, A.C.J., **3**.

Barbosa, R.S. et al. (2004) Dynamics of the fractional-order Van der Pol oscillator, *Iccc 2004: Second Ieee International Conference on Computational Cybernetics, Proceedings*, 373-378.

Barbosa, R.S. et al. (2007) Analysis of the van der pol oscillator containing derivatives of fractional order, *J Vib Control* **13**, 1291-1301.

Baz, A. & Ro, J. (1996) Vibration control of plates with active constrained layer damping, *Smart Materials & Structures* **5**, 272-280.

Bhalekar, S. & Daftardar-Gejji, V. (2010) Fractional ordered Liu system with time-delay, *Commun Nonlinear Sci* **15**, 2178-2191.

Bhalekar, S. & Gejji, V.D. (2011) A predictor-corrector scheme for solving nonlinear delay differential equations of fractional order, *Journal of Fractional Calculus and Applications* **1:5** 1-9.

Bhalekar, S. et al. (2012) Genalized fractional order Bloch equation with extended delay, *Int J Bifurcat Chaos* **22: 4**, 1250071.

Bi, Q. & Dai, H.H. (2000) Analysis of non-linear dynamics and bifurcations of a shallow arch subjected to periodic excitation with internal resonance, *J Sound Vib* **233**, 557-571.

Blandford, G.E. (1997) Review of Progressive failure analyses for truss structures, *Journal of Structure and Engineering* **123**, 122-129.

Bonnet, C. & Partington, J.R. (2007) Stabilization of some fractional delay systems of neutral type, *Automatica* **43**, 2047-2053.

Cao, J.Y. et al. (2010) Nonlinear dynamics of Duffing system with fractional order damping, *J Comput Nonlin Dyn* **5**.

Caponetto, R. et al. (2010) *Fractional order systems: modeling and control applications*, World Scientific Series on Nonlinear Science, Series A (World Scientific, Singapore).

Caputo, M. (1967) Linear models of dissipation whose Q is almost frequency independent–II, *Geophys. J. Roy. Astron. Soc.* **13**, 529-539.

Caputo, M. (1974) Vibrations of an infinite viscoelastic layer with a dissipative memory, *J Acoust Soc Am* **56**, 897-904.

Caputo, M. & Mainardi, F. (1971a) Linear models of dissipation in anelastic solids, *Rivista del Nuovo Cimento (Ser. II)* **1**, 161-198.

Caputo, M. & Mainardi, F. (1971b) A new dissipation model based on memory mechanism, *Pure Applied Geophys* **91:1**, 134-147.

Cartmell, M.P. (1992) *An introduction to linear, parametric and nonlinear vibrations* (Chapman & Hall, London).

Casal, A. & Freedman, M. (1980) A Poincaré-Lindstedt approach to bifurcation problems for differential-delay equations, *IEEE Transactions on Automatic Control* **25**, 967-973.

Cederbaum, G. & Mond, M. (1994) Instability and chaos in the elastica type problem of parametrically excited columns, *J Sound Vib* **176:4**, 475-486.

Chen, G.R. et al. (2000) Bifurcation control: Theories, methods, and applications, *Int J Bifurcat Chaos* **10**, 511-548.

Chen, J.H. & Chen, W.C. (2008) Chaotic dynamics of the fractionally damped van der Pol equation, *Chaos Soliton Fract* **35**, 188-198.

Chen, J.S. & Yang, M.R. (2007) Vibration and stability of a shallow arch under a moving mass-dashpot-spring system, *J Vib Acoust* **129**, 66-72.

Chen, L.Q. & Cheng, C.J. (2000) Dynamical behavior of nonlinear viscoelastic columns based on 2-order Galerkin truncation, *Mech Res Commun* **27:4**, 413-419.

Chen, L.Q. et al. (2004) Transient responses of an axially accelerating viscoelastic string constituted by a fractional differentiation law, *J Sound Vib* **278**, 861-871.

Chen, W.C. (2008) Nonlinear dynamics and chaos in a fractional-order financial system, *Chaos Soliton Fract* **36**, 1305-1314.

Chen, Y.Q. et al. (2006) Robust stability check of fractional order linear time invariant systems with interval uncertainties, *Signal Process* **86**, 2611-2618.

Chen, Y.Q. & Moore, K. (2002) Analytical Stability bound for a class of delayed fractional order dynamic systems, *Nonlinear Dynam*, 191-200.

Chen, Y.Q. & Moore, K.L. (2002) Analytical stability bound for a class of delayed fractional-order dynamic systems, *Nonlinear Dynam* **29**, 191-200.

Cichon, C. & Corradi, L. (1981) Large displacements analysis of elastic-plastic trusses with unstable bars, *Engineering Structures* **3**, 210-218.

Craiem, D. & Magin, R.L. (2010) Fractional order models of viscoelasticity as an alternative in the analysis of red blood cell (RBC) membrane mechanics, **Phys Biol 7**.

Craiem, D. et al. (2008) Fractional-order viscoelasticity applied to describe uniaxial stress relaxation of human arteries, *Phys Med Biol* **53**, 4543-4554.

Cuvalci, O. & Ertas, A. (1996) Pendulum as vibration absorber for flexible structures: Experiments and theory, *J Vib Acoust* **118**, 558-566.

Daftardar-Gejji, V. et al. (2012) Dynamics of fractional-ordered Chen system with delay, *Pramana—J Phys* **79**, 61-69.

Danca, M.F. (2010) Chaotic behavior of a class of discontinuous dynamical systems of fractional-order, *Nonlinear Dynam* **60**, 525-534.

Debnath, L. (2003) "Recent applications of fractional calculus to science and engineering," *Int. J. Math. Math. Sci*, **2003**, 3413-3442.

De-Freitas, J.A.T. & Ribeiro, A.C.B.S. (1992) Large displacement elastoplastic analysis of space trusses, *Computers and Structures* **44**, 1007-1016.

Delbosco, D. & Rodino, L. (1996) Existence and uniqueness for a nonlinear fractional differential equation, *J Math Anal Appl* **204**, 609-625.

Demir, D.D. et al. (2012) Application of fractional calculus in the dynamics of beams, *Bound Value Probl.*

Deng, W.H. (2007) Short memory principle and a predictor-corrector approach for fractional differential equations, *J Comput Appl Math* **206**, 174-188.

Deng, W.H. et al. (2007) Stability analysis of linear fractional differential system with multiple time delays, *Nonlinear Dynam* **48**, 409-416.

Deng, W.H. et al. (2007) Stability analysis of linear fractional differential system with multiple time delays, *Nonlinear Dynam* **48**, 409-416.

Deü, J.F. et al. (2004). Dynamic responses of flexible mechanisms with viscoelastic constrained layer damping treatment using a fractional derivative model, *Proc. European Congress on Computational Methods in Applied Sciences and Engineering, ECCOMAS 2004*, ed. Neittaanmaki, P. et al., pp. 1-18.

Deü, J.F. et al. (2008) Dynamic responses of flexible-link mechanisms with passive/active damping treatment, *Computers & Structures* **86**, 258-265.

Di Paola, M. et al. (2011) Visco-elastic behavior through fractional calculus: An easier method for best fitting experimental results, *Mech Mater* **43**, 799-806.

Diethelm, K. et al. (2002) A predictor-corrector approach for the numerical solution of fractional differential equations, *Nonlinear Dynam* **29**, 3-22.

Diethelm, K. et al. (2004) Detailed error analysis for a fractional Adams method, *Numer Algorithms* **36**, 31-52.

Diethelm, K. et al. (2005) Algorithms for the fractional calculus: A selection of numerical methods, *Computer Methods in Applied Mechanics and Engineering* **194**, 743-773.

Djordjevic, V.D. et al. (2003) Fractional derivatives embody essential features of cell rheological behavior, *Ann Biomed Eng* **31**, 692-699.

Draganescu, G.E. (2006) Application of a variational iteration method to linear and nonlinear viscoelastic models with fractional derivatives, *J Math Phys* **47**.

Drozdov, A.D. (1997) Fractional differential models in finite viscoelasticity, *Acta Mech* **124**, 155-180.

Eissa, M. & Amer, Y.A. (2004) Vibration control of a cantilever beam subject to both external and parametric excitation, *Appl Math Comput* **152**, 611-619.

El-Bassiouny, A.F. (2006) Single-mode control and chaos of cantilever beam under primary and principal parametric excitations, *Chaos Soliton Fract* **30**, 1098-1121.

EI-Bassiouny, A.F. (2005) Effect of nonlinearities in elastomeric material dampers on torsional oscillation control, *Appl. Math. Comput.***162**: 835-854.

Elsayed, A.M.A. (1992) On the fractional differential-equations, *Appl Math Comput* **49**, 205-213.

Enelund, M. & Josefson, B.L. (1997) Time-domain finite element analysis of viscoelastic structures with fractional derivatives constitutive relations, *AIAA Journal* **35**, 1630-1637.

Ferry, J.D. (1980) Viscoelastic properties of polymers (3rd Edition ed.)(Wiley, New York).

Francesco, M. (2010) Fractional calculus and waves in linear viscoelasticity: an *introduction to mathematical models* (Imperial College Press, London).

Fukunaga, M. & Shimizu, N. (2009). Nonlinear fractional derivative stress-strain relations for polymer gels based on the generalized Maxwell model, *Proc. ASME 2009 International Design Engineering Technical Conferences & Computers and Information in Engineering Conference*, pp. 1-6.

Fukunaga, M. et al. (2009) A nonlinear fractional derivative model of impulse motion for viscoelastic materials, *Phys Scripta* **T136**, 014010-6.

Fuller, C.R. et al. (1997) *Active Control of Vibration* (Academic Press, London).

Fuller, C.R. et al. (1997) *Active Control of Vibration* (Academic, London).

Gafiychuk, V. & Datsko, B. (2008) Stability analysis and limit cycle in fractional system with Brusselator nonlinearities, *Phys Lett A* 372, 4902-4904.

Galucio, A.C. et al. (2004) Finite element formulation of viscoelastic sandwich beams using fractional derivative operators, *Comput Mech* **33**, 282-291.

Galucio, A.C. et al. (2007) Hybrid active-passive damping treatment of sandwich beams in non-linear dynamics, *J Vib Control* **13**, 851-881.

Ge, Z.M. & Hsu, M.Y. (2007) Chaos in a generalized van der Pol system and in its fractional order system, *Chaos Soliton Fract* **33**, 1711-1745.

Ge, Z.M. & Jhuang, W.R. (2007) Chaos, control and synchronization of a fractional order rotational mechanical system with a centrifugal governor, *Chaos Soliton Fract* **33**, 270-289.

Ge, Z.M. & Ou, C.Y. (2007) Chaos in a fractional order modified Duffing system, *Chaos Soliton Fract* **34**, 262-291.

Ge, Z.M. & Zhang, A.R. (2007) Chaos in a modified van der Pol system and in its fractional order systenis, *Chaos Soliton Fract* **32**, 1791-1822.

Gemant, A. (1936) A method of analyzing experimental results obtained from elastic-viscous bodies, *Physics* (*New York*) **7**, 311-317.

Gil-Negrete, N. et al. (2009) A Nonlinear rubber material model combining fractional order viscoelasticity and amplitude dependent Effects, *J Appl Mech–T Asme* **76**.

Gjurchinovski, A. et al. (2010) Delayed feedback control of fractional-order chaotic systems, *J Phys a–Math Theor* **43**.

Grigorenko, I. & Grigorenko, E. (2003) Chaotic dynamics of the fractional Lorenz system, *Phys Rev Lett* **91**.

Guo, Z.J. (2012) *Residue harmonic balance for steady state responses of nonlinear dynamical systems*. PhD Thesis, City university of hong kong.

Guo, Z.J. et al. (2011) Oscillatory region and asymptotic solution of fractional van der Pol oscillator via residue harmonic balance technique, *Appl Math Model* **35**, 3918-3925.

Guo, Z.J., Leung, A.Y.T. & Ma, X.Y. (2014) Solution procedure of residue harmonic balance method and its applications, *Sci. China Phys. Mech.* **57**, 1581-1591.

Haddad, Y.M. (1995) *viscoelasticity of engineering materials* (Kluwer Academic Publishers, Dordrecht).

Hall, K.C. et al. (2000) Proper orthogonal decomposition technique for transonic unsteady aerodynamic flows, *AIAA Journal* **38:10**, 1853-1862.

Hartley, T.T. & Lorenzo, C.F. (2002) Dynamics and control of initialized fractional-order systems, *Nonlinear Dynam* **29**, 201-233.

Hartley, T.T. et al. (1995) Chaos in a fractional order Chua's system, *Ieee T Circuits–I* **42**, 485-490.

Hashim, I. et al. (2009) Homotopy analysis method for fractional IVPs, *Commun Nonlinear Sci* **14**, 674-684.

He, J.H. (1998) Approximate analytical solution for seepage flow with fractional derivatives in porous media, *Computer Methods in Applied Mechanics and Engineering* **167**, 57-68.

Hedrih, K. (2008) Vibration modes of an axially moving double belt system with creep layer, *J Vib Control* **14**, 1333-1347.

Hilfer, R. (2000) *Applications of fractional calculus in physics* (World Scientific, Singapore).

Hill, C. et al. (1989) Post-buckling analysis of steel space trusses, *Journal of Structural Engineering* **115**, 900-918.

Hu, H.Y. (2001). Abundant dynamic features of a non-linear system under delayed feedback control, *Proc. Proceedings of the 6th Asia Pacific vibration conference*, **1**, pp. 11-15.

Hu, H.Y. et al. (1998) Resonances of a harmonically forced duffing oscillator with time delay state feedback, *Nonlinear Dynam* **15**, 311-327.

Hu, H.Y. & Wang, Z.H. (2002) *Dynamics of controlled mechanical systems with delayed feedback* (Springer Verlag, Berlin).

Hu, S. et al. (2012) *Fractional processes and fractional–order signal processing* (Springer-Verlag, London).

Hwang, C. & Cheng, Y.C. (2006) A numerical algorithm for stability testing of fractional delay systerns, *Automatica* **42**, 825-831.

Hwang, J.S. & Ku, S.W. (1997) Analytical modeling of high damping rubber bearings, *J*

Struct Eng–Asce **123**, 1029-1036.

Jordanov, I.N. & Cheshankov, B.I. (1988) Optimal design of linear and non-linear dynamic vibration absorbers, *J. Sound Vib.***123**, 157-170.

Jones, D.I.G. (2001) *Handbook of viscoelastic vibration damping* (John Wiley and Sons, England).

Ji, J.C. (2001) Local bifurcation control of a forced single-degree-of-freedom nonlinear system: Saddle-node bifurcation, *Nonlinear Dynam* **25**, 369-382.

Ji, J.C. & Hansen, C.H. (2006) Stability and dynamics of a controlled van der Pol-Duffing oscillator, *Chaos, Solitons and Fractals* **28**, 555-570.

Ji, J.C. & Leung, A.Y.T. (2002) Bifurcation control of a parametrically excited duffing system, *Nonlinear Dynam* **27**, 411-417.

Ji, J.C. & Leung, A.Y.T. (2002) Resonances of a non-linear s.d.o.f. system with two time delays in linear feedback control, *J Sound Vib* **253:5**, 985-1000.

Kassimali, A. & Bidhendi, E. (1988) Stability of Trusses under Dynamic Loads, *Computers & Structures* **29**, 381-392.

Katsikadelis, J.T. (1994) The analog equation method-a powerful bem-based solution technique for solving linear and nonlinear engineering problems, *Boundary Element Method Xvi*, 167-182.

Katsikadelis, J.T. (2009) Numerical solution of multi-term fractional differential equations, *Zamm-Z Angew Math Me* **89**, 593-608.

Katsikadelis, J.T. (2011) The BEM for numerical solution of partial fractional differential equations, *Comput Math Appl* **62**, 891-901.

Katsikadelis, J.T. (2012) Nonlinear dynamic analysis of viscoelastic membranes described with fractional differential models, *J Theor App Mech–Pol* **50**, 743-753.

Katsikadelis, J.T. & Babouskos, N.G. (2010) Post-buckling analysis of viscoelastic plates with fractional derivative models, *Eng Anal Bound Elem* **34**, 1038-1048.

Koeller, R.C. (1984) Applications of fractional calculus to the theory of viscoelasticity, *J Appl Mech–T Asme* **51**, 299-307.

Koeller, R.C. (1986) Polynomial operators, stieltjes convolution, and fractional calculus in hereditary mechanics, *Acta Mech* **58**, 251-264.

Koh, C.G. & Kelly, J.M. (1990) Application of fractional derivatives to seismic analysis of base-Isolated models, *Earthquake Eng Struc* **19**, 229-241.

Kondoh, K. & Atluri, S.N. (1985) Influence of local buckling on global instability: simplified, large deformation, post-buckling analyses of plane trusses, *Computers and Structures* **21**, 613-27.

Kilbas, A.A., Srivastava, H.M. & Trujillo, J.J. (2006) *Theory and applications of fractional differential equations*, Elsevier, New York.

Kryloff, N. & Bogoliuboff, N. (1947) *Introduction to nonlinear mechanics* (Princeton University Press, Princeton).

Kuang, Y. (1993) *Delay differential equations with applications in population dynamics* (Academic Press, Boston).

Kulish, V.V. & Lage, J.L. (2002) Application of fractional calculus to fluid mechanics, *J Fluid Eng-T Asme* **124**, 803-806.

Lai, W.M. et al. (2010) *Introduction to continuum mechanics (4th Edition)*.

Laskin, N. (2000) Fractional quantum mechanics, *Phys. Rev. E***62**, 3135-3145.

Lau, S.L. & Cheung, Y.K. (1981) Amplitude incremental variational principle for nonlinear vibration of elastic systems, *ASME Journal of Applied Mechanics* **48**, 959-964.

Lepik, U. (2009) Solving fractional integral equations by the Haar wavelet method, *Appl Math Comput* **214**, 468-478.

Leu, L.J. & Yang, Y.B. (1990) Effects of rigid body and stretching on nonlinear analysis of trusses, *Journal of Structure Engineering* **116**, 2582-2598.

Leung, A.Y.T. & Fung, T.C. (1990) Nonlinear Steady-State Vibration and Dynamic Snap through of Shallow Arch Beams, *Earthquake Eng Struc* **19**, 409-430.

Leung, A.Y.T. & Guo, Z.J. (2011) Forward residue harmonic balance for autonomous and non-autonomous systems with fractional derivative damping, *Commun Nonlinear Sci* **16**, 2169-2183.

Leung, A.Y.T. & Guo, Z.J. (2011a) Residue harmonic balance approach to limit cycles of nonlinear jerk equations, *Int J Nonlin Mech* **46:6**, 898-906.

Leung, A.Y.T. & Guo, Z.J. (2011b) Residue harmonic balance for two-degree-of-freedom airfoils with cubic structural nonlinearity, *AIAA Journal* **49**, 2607-2615.

Leung, A.Y.T. et al. (2012) Steady state bifurcation of a periodically excited system under delayed feedback controls, *Commun Nonlinear Sci* **17**, 5256-5272.

Leung, A.Y.T. et al. (2012) Steady state bifurcation of a periodically excited system under delayed feedback controls, *Commun Nonlinear Sci* **17**, 5256-5272.

Leung, A.Y.T. et al. (2012) Transition curves and bifurcations of a class of fractional Mathieu-type equations, *Int J Bifurcat Chaos* **22**.

Leung, A.Y.T. et al. (2012) Transition curves and bifurcations of a class of fractional Mathieu-type equations, *Int J Bifurcat Chaos* **22**, 1250275-13.

Leung, A.Y.T. et al. (2004) Resonance control for a forced single-degree-of-freedom nonlinear system, *Int J Bifurcat Chaos* **14**, 1423-1429.

Leung, A.Y.T. et al. (2012) The residue harmonic balance for fractional order van der Pol like oscillators, *J Sound Vib* **331**, 1115-1126.

Leung, A.Y.T. et al. (2013) Steady state response of fractionally damped nonlinear viscoelastic arches by residue harmonic homotopy, *Computers & Structures* **121**, 10-21.

Leung, A.Y.T. & Guo, Z.J. (2013) Effects of fractional differentiation orders on a coupled Duffing circuit, *Int. J. Bifurcation and Chaos* **23**, 1350193.

Leung, A.Y.T. & Zhang, Q.C. (1998) Higher order normal form and period averaging, *J Sound Vib* **217**, 795-806.

Li, C.G. & Chen, G.R. (2004) Chaos and hyperchaos in the fractional-order Rössler equations, *Physica A* **341**, 55-61.

Li, C.P. & Wang, Y.H. (2009) Numerical algorithm based on Adomian decomposition for fractional differential equations., *Comput Math Appl* **57**, 1672-1681.

Li, G.G. (2002) Dynamical behaviours of Timoshenko beam with fractional derivative constitutive relation, *International Journal of Nonlinear Science and Numerical*

Simulations **3**, 67-73.

Li, G.G. et al. (2001) Dynamical stability of viscoelastic column with fractional derivative constitutive relation, *Appl Math Mech–Engl* **22**, 294-303.

Li, G.G. et al. (2003) Application of Galerkin method to dynamical behavior of viscoelastic timoshenko beam with finite deformation, *Mech Time–Depend Mat* **7**, 175-188.

Li, J. et al. (2007) Cubic velocity feedback control of high-amplitude vibration of a nonlinear plant to a primary resonance excitation, *Shock Vib* **14**, 1-14.

Li, T.Y. (1997) Numerical solution of multivariate polynomial systems by homotopy continuation methods, *Acta Numerica* **6**, 399-436.

Li, X.Y. et al. (2006) Response of parametrically excited Duffing-van der Pol oscillator with delayed feedback, *Applied Mathematics and Mechanics (English Edition)* **27:12**, 1585-1595.

Li, X.Y. et al. (2006) The response of a Duffing-van der Pol oscillator under delayed feedback control, *J Sound Vib* **291**, 644-655.

Li, Y. et al. (2009) Mittag-Leffler stability of fractional order nonlinear dynamic systems, *Automatica* **45**, 1965-1969.

Lim, C.W. & Wu, B.S. (2003) A new analytical approach to the Duffing-harmonic oscillator, *Phys Lett A* **311**, 365-373.

Lion, A. (1997) On the thermodynamics of fractional damping elements, *Continuum Mech Therm* **9**, 83-96.

Liu, J.G. & Xu, M.Y. (2008). Study on the viscoelasticity of cancellous bone based on higher-order fractional models, *Proc. The 2nd International Conference on Bioinformatics and Biomedical Engineering*, pp. 1733-1736.

Lu, J.G. (2006) Chaotic dynamics of the fractional-order Ikeda delay system and its synchronization, *Chinese Phys* **15**, 301-305.

Lu, J.G. & Chen, G.R. (2006) A note on the fractional-order Chen system, *Chaos Soliton Fract* **27**, 685-688.

Luo, A.C.J. (2014) On analytical routes to chaos in nonlinear systems, *Int. J. Bifurcation and Chaos* **24**, 1430013.

Ma, S.C. et al. (2012) Numerical solutions of a variable-order fractional financial system, J Appl Math.

Maccari, A. (2001) The response of a parametrically excited van der Pol oscillator to a time delay state feedback, *Nonlinear Dynam* **26**, 105-119.

Maccari, A. (2008) Vibration amplitude control for a van der Pol-Duffing oscillator with time delay, *J Sound Vib* **317**, 20-29.

Maccari, A. (2012) Response control for the externally excited van der Pol oscillator with non-local feedback, *J Sound Vib* **331**, 987-995.

MacDonald, N. (1989) *Biological delay systems: Linear stability theory* (Cambridge University, New York).

Macdonald, N. (1995) Harmonic balance in delaye-differential equations, *J Sound Vib* **186:4**, 649-656.

Magin, R. et al. (2011) On the fractional signals and systems, *Signal Process* **91**, 350-371.

Magin, R.L. (2006) *Fractional Calculus in Bioengineering* (Begell House Publishers, Redding).

Magin, R.L. (2010) Fractional calculus models of complex dynamics in biological tissues, *Comput Math Appl* **59**, 1586-1593.

Magin, R.L. et al. (2008) Anomalous diffusion expressed through fractional order differential operators in the Bloch-Torrey equation, *Journal of Magnetic Resonance* **190:2**, 255-270.

Mainardi, F. (1996) Fractional relaxation-oscillation and fractional diffusion-wave phenomena, *Chaos Soliton Fract* **7**, 1461-1477.

Makris, N. (1997) Three-dimensional constitutive viscoelastic laws with fractional order time derivatives, *Journal of Rheology* **41**, 1007-1020.

Makris, N. & Constantinou, M.C. (1991) Fractional-derivative Maxwell model for viscous dampers, *J Struct Eng–Asce* **117**, 2708-2724.

Makroglou, A. et al. (1994) Computational results for a feedback control for a rotating viscoelastic beam, *J Guid Control Dynam* **17**, 84-90.

Malhotra, N. & Namachchivaya, N.S. (1997) Chaotic dynamics of shallow arch structures under 1:2 resonance, *J Eng Mech–Asce* **123**, 612-619.

Maraaba, T. et al. (2008) Existence and uniqueness theorem for a class of delay differential equations with left and right Caputo fractional derivatives, *J Math Phys* **49**.

Matignon, D. (1996). Stability results for fractional differential equations with applications to control processing *Proc. Computational Engineering in Systems and Application Multiconference, IMACS, IEEE–SMC*, **2**, pp. 963-968.

Meador, D.J. & Mead, D.J. (1999) *Passive Vibration Control* (Wiley, New York).

Meerschaert, M.M. & Sikorskii, A. (2012) *Stochastic models for fractional calculus* (De Gruyter, Berlin: Boston).

Meral, F.C. et al. (2010) Fractional calculus in viscoelasticity: An experimental study, *Commun Nonlinear Sci* **15**, 939-945.

Miller, K.S. & Ross, B. (1993) *An introduction to the fractional calculus and fractional differential equations* (John Wiley & Sons, New York).

Monje, C.A. et al. (2010) *Fractional order systems and controls: Fundamentals and applications* (Springer-Verlag, London).

Mook, D.T. et al. (1986) The Influence of an Internal Resonance on Nonlinear Structural Vibrations under Combination Resonance Conditions, *J Sound Vib* **104**, 229-241.

Morrison, T.M. & Rand, R.H. (2007) 2:1 Resonance in the delayed nonlinear Mathieu equation, *Nonlinear Dynam* **50**, 341-352.

Müller, S. et al. (2011) A nonlinear fractional viscoelastic material model for polymers, *Comp Mater Sci* **50**, 2938-2949.

Nakra, B.C. (1998) Vibration control in machines and structures using viscoelastic damping, *J Sound Vib* **211**, 449-465.

Nashif, A.D. et al. (1985) *Vibration damping* (John Willey & Sons, New York).

Nasuno, H. & Shimizu, N. (2008) Power time numerical integration algorithm for nonlinear fractional differential equations, *J Vib Control* **14**, 1313-1332.

Nasuno, H. et al. (2007). Fractional derivative consideration on nonlinear viscoelastic stoical and dynamical behaviour under large ore-displacement, *Proc. Advances in Fractional Calculus: Theoretical Developments and Applications in Physics and Engineering*, ed. al., J.S.e., pp. 363-372.

Nasuno, H. et al. (2006) Geometrical Nonlinear Statical and Dynamical Models of Fractional Derivative Viscoelastic Body, *Transactions of the Japan Society of Mechanical Engineers*. C **72:76**, 1041-1048.

Natsiavas, S. (1992) Steady state oscillations and stability of nonlinear dynamic vibration absorbers, *J. Sound Vib.***156**, 227-245.

Nayfeh, A.H. & Mook, D.T. (1979) *Nonlinear Oscillations* (Wiley-Interscience, New York).

Nerantzaki, M.S. & Babouskos, N.G. (2011) Analysis of inhomogeneous anisotropic viscoelastic bodies described by multi-parameter fractional differential constitutive models, *Comput Math Appl* **62**, 945-960.

Nutting, P.G. (1921) A new general law of deformation, *J Franklin I* **191**, 678-685.

Odibat, Z.M. & Momani, S. (2006) Application of variational iteration method to Nonlinear differential equations of fractional order, *Int J Nonlin Sci Num* **7**, 27-34.

Oldham, K.B. & Spanier, J. (1974) *The fractional calculus* (Academic Press, New York).

Osler, T.J. (1970) The fractional derivative of a composite function, *Siam J Math Anal* **1:2**, 288-293.

Oueini, S.S. & Nayfeh, A.H. (1999) Single-mode control of a cantilever beam under principal parametric excitation, *J Sound Vib* **224**, 33-47.

Oueini, S.S. et al. (1998) A nonlinear vibration absorber for flexible structures, *Nonlinear Dynam* **15**, 259-282.

Padovan, J. (1987) Computational algorithms for fe formulations involving fractional operators, *Comput Mech* **27**, 271-287.

Padovan, J. et al. (1987) Asymptotic Steady-State Behavior of Fractionally Damped Systems, *J Franklin I* **324**, 491-511.

Padovan, J. & Sawicki, J.T. (1998) Nonlinear vibrations of fractionally damped systems, *Nonlinear Dynam* **16**, 321-336.

Pankaj, W. & Anindya, C. (2004) Averaging oscillations with small fractional damping and delayed terms, *Nonlinear Dynam* **38**, 3-22.

Papadrakakis, M. (1983) Inelastic post-buckling analysis of trusses, *Journal of Structural Engineering* **109**, 2129-2147.

Papoulia, K.D. & Kelly, J.M. (1997) Visco-hyperelastic model for filled rubbers used in vibration isolation, *J Eng Mater–T Asme* **119**, 292-297.

Peng, Z.K., Lang, Z.Q., Billings, S.A. & Tomlinson, G.R. (2008) Comparisons between harmonic balance and nonlinear output frequency response function in nonlinear system analysis, *J. Sound Vib.***311**, 56-73.

Perko, L.M. (1968) Higher order averaging and related methods for perturbed periodic and quasi-periodic systems, *SIAM Journal of Applied Mathematics* **17:4**, 469-724.

Petráš, I. (2011) *Fractional–order nonlinear systems: Modeling, analysis and simulation* (Springer Verlag, London).

Plaut, R.H. & Hsieh, J.C. (1985) Oscillations and Instability of a Shallow Arch under 2-Frequency Excitation, *J Sound Vib* **102**, 189-201.

Plaut, R.H. & Hsieh, J.C. (1987a) Chaos in a mechanism with time delays under parametric and external excitation, *J Sound Vib* **114**, 73-90.

Plaut, R.H. & Hsieh, J.C. (1987b) Nonlinear structural vibrations involving a time delay in damping, *J Sound Vib* **117**, 497-510.

Podlubny, I. (1999a) Fractional-order systems and $PI^{\lambda}D^{\mu}$-controllers, *IEEE Trans. Automatic Control* **44**, 208-214.

Podlubny, I. (1999b) *Fractional differential equations: An introduction to fractional derivatives, fractional differential equations to methods of their solutions and some of their applications* (Academic Press, New York).

Pratiher, B. (2012) Vibration control of a transversely excited cantilever beam with tip mass, *Archive of Applied Mechanics* **82**, 31-42.

Pritz, T. (1996) Analysis of four-parameter fractional derivative model of real solid materials, *J Sound Vib* **195**, 103-115.

Pritz, T. (2003) Five-parameter fractional derivative model for polymeric damping materials, *J Sound Vib* **265**, 935-952.

Pyragas, K. (1992) Continuous control of chaos by self-controlling feedback, *Phys Lett A*, 421-428.

Qian, C.Z. & Tang, J.S. (2008) A time delay control for a nonlinear dynamic beam under moving load, *J Sound Vib* **309**, 1-8.

Ramrakhyani, D.S. et al. (2004) Modeling of elastomeric materials using nonlinear fractional derivative and continuously yielding friction elements, *Int J Solids Struct* **41**, 3929-3948.

Rand, R.H. et al. (2010) Fractional Mathieu equation, *Commun Nonlinear Sci* **15**, 3254-3262.

Reddy, J.N. (2004a) *An introduction to nonlinear finite element analysis* (Oxford University Press, London).

Reddy, J.N. (2004b) *Mechanics of Laminated composite plates and shells theory and analysis* (CRC Press, Boca RatonFla, USA).

Richard, M. et al., (2011) On the fractional signals and systems, *Signal Process.***91**, 350-371.

Rossikhin, Y.A. & Shitikova, M.V. (1997) Applications of fractional calculus to dynamic problems of linear and nonlinear hereditary mechanics of solids, *Applied Mechanics Reviews* **50**, 15-67.

Rossikhin, Y.A. & Shitikova, M.V. (1998) Application of fractional calculus for analysis of nonlinear damped vibrations of suspension bridges, *J Eng Mech−Asce* **124**, 1029-1036.

Rossikhin, Y.A. & Shitikova, M.V. (2000) Analysis of nonlinear vibrations of a two-degree-of-freedom mechanical system with damping modelled by a fractional derivative, *Journal of Engineering Mathematics* **37**, 343-362.

Rossikhin, Y.A. & Shitikova, M.V. (2003) Free damped nonlinear vibrations of a viscoelastic plate under two-to-one internal resonance, *Mater Sci Forum* **440-4**, 29-36.

Rossikhin, Y.A. & Shitikova, M.V. (2006) Analysis of free non-linear vibrations of a viscoelastic plate under the conditions of different internal resonances, *Int J Nonlin Mech* **41**, 313-325.

Rossikhin, Y.A. & Shitikova, M.V. (2009) New approach for the analysis of damped vibrations of fractional oscillators, *Shock Vib* **16**, 365-387.

Rossikhin, Y.A. & Shitikova, M.V. (2010) Application of fractional calculus for dynamic problems of solid mechanics: Novel trends and recent results, *Applied Mechanics Reviews* **63**, 010801-52.

Rossikhin, Y.A. & Shitikova, M.V. (2010) Application of Fractional Calculus for Dynamic Problems of Solid Mechanics: Novel Trends and Recent Results, Applied *Mechanics Reviews* **63**.

Rostami, M. & Haeri, M. (2013) Study of limit cycles and stability analysis of fractional Arneodo oscillator, *J Optimiz Theory App* **156**, 68-78.

Ryabov, Y.E. & Puzenko, A. (2002) Damped oscillations in view of the fractional oscillator equation, *Physical Review B* **66**.

Sabatier, J. et al. (2007) *Advances in fractional calculus: Theoretical developments and applications in physics and engineering* (SpringerVerlag, Dordrecht).

Saenz, A.W. (1991) Higher-order averaging for nonperiodic systems, *J Math Phys* **32**, 2679-2694.

Saichev, A.I. & Zaslavsky, G.M. (1997) Fractional kinetic equations: solutions and applications, *Chaos*, **7**, 753-764.

Samko, S.G. et al. (1993) *Fractional Integrals and Derivatives: Theory and Applications* (Gordon and Breach, Amsterdam).

Sardar, T. et al. (2009) The analytical approximate solution of the multi-term fractionally damped Van der Pol equation, *Phys Scripta* **80**.

Sayed, M. & Kamel, M. (2012) 1:2 and 1:3 internal resonance active absorber for non-linear vibrating system, *Appl Math Model* **36**, 310-332.

Sayed, M., Hamed, Y.S. & Amer, Y.A. (2011) Vibration reduction and stability of nonlinear system subjected to external and parametric excitation forces under a nonlinear absorber, Int. J. *Contemp. Math. Sci.***6**, 1051-1070.

Scalas, E. et al. (2000) Fractional calculus and continuous-time finance, *Physica A* **284**, 376-384.

Schiessel, H. et al. (1995) Generalized viscoelastic models: Their fractional equations with solutions, *J Phys a−Math Gen* **28**, 6567-6584.

Schmidt, A. & Gaul, L. (2002) Finite element formulation of viscoelastic constitutive equations using fractional time derivatives, *Nonlinear Dynam* **29**, 37-55.

Scott-Blair, G.W. (1947) The role of psychophysics in rheology, *Journal of Colloid Scieaces* **2**, 21-32.

Sekikawa, M. et al. (2004) Bifurcation structure of fractional harmonic entrainments in the forced Rayleigh oscillator, *Electronics and Communications in Japan* (*Part III*: *Fundamental Electronic Science*) **87**, 30-40.

Shawagfch, N.T. (2002) Analytical approximate solutions for nonlinear fractional differential equations, *Appl Math Comput* **131**, 517-529.

Shen, Y.J. et al. (2012) Primary resonance of Duffing oscillator with fractional-order derivative, *Commun Nonlinear Sci* **17**, 3092-3100.

Shen, Y.J. et al. (2012) Primary resonance of Duffing oscillator with two kinds of fractional-order derivatives, *Int J Nonlin Mech* **47**, 975-983.

Sheu, L.J. et al. (2007) Chaotic dynamics of the fractionally damped Duffing equation, *Chaos Soliton Fract* **32**, 1459-1468.

Sheu, L.J. et al. (2008) Chaos in the Newton-Leipnik system with fractional order, *Chaos Soliton Fract* **36**, 98-103.

Shimizu, N. & Zhang, W. (1999) Fractional calculus approach to dynamic problems of viscoelastic materials, JSME international journal. *Series C, Mechanical systems, machine elements and manufacturing* **42**, 825-837.

Siewe, M.S. et al. (2012) Parametric resonance in the Rayleigh-Duffing oscillator with time-delayed feedback, *Commun Nonlinear Sci* **17**, 4485-4493.

Sjöberg, M. & Kari, L. (2002) Non-linear behavior of a rubber isolator system using fractional derivatives, *Vehicle Syst Dyn* **37**, 217-236.

Sjöberg, M. & Kari, L. (2003) Nonlinear isolator dynamics at finite deformations: An effective hyperelastic, fractional derivative, generalized friction model, *Nonlinear Dynam* **33**, 323-336.

Skaar, S.B. et al. (1988) Stability of viscoelastic control systems, *IEEE Transactions on Automatic Control* **33**, 348-357.

Smith, E.M. (1994) Nonlinear analysis of space trusses, *Journal of Structural Engineering* **120**, 2717-2736.

Smith, M.E. (1994) Compression member models for space trusses: review, *Journal of Structural Engineering* **120**, 2399-2470.

Sommese, A.J. & Wampler, C.W. (2005) *The numerical solution of systems of polynomials: arising in engineering and science* (World Scientific, Singapore).

Song, Y., Sato, H., Iwata, Y. & Komatsuzaki, T. (2003) The response of a dynamic vibration absorber system with a parametrically excited pendulum, *J. Sound Vib.* **259**, 747-759.

Soom, A. & Lee, M. (1983) Optimal design of linear and nonlinear vibration absorbers for damped system, *J. Vib. Acoust.***105**, 112-119.

Soong, T.T. & Dargush, G.F. (1997) *Passive energy dissipation systems in structural engineering* (Wiley, London).

Stevens, K.K. (1969) Transverse vibration of a viscoelastic column with initial curvature under periodic axial load, *J Appl Mech—T Asme* **36:4**, 814-818.

Stevens, K.K. (1996) On the parametric excitation of a viscoelastic column, *AIAA Journal* **4:12**, 2111-2116.

Stiassnie, M. (1979) Application of fractional calculus for the formulation of viscoelastic models, *Appl Math Model* **3**, 300-302.

Suarez, L.E. & Shokooh, A. (1997) An eigenvector expansion method for the solution of motion containing fractional derivatives, *J Appl Mech—T Asme* **64**, 629-635.

Suchorsky, M.K. & Rand, R.H. (2012) A pair of van der Pol oscillators coupled by fractional derivatives, *Nonlinear Dynam* **69**, 313-324.

Suki, B. et al. (1994) Lung-tissue viscoelasticity -a mathematical framework and its molecular-basis, *J Appl Physiol* **76**, 2749-2759.

Suvorova, Y.V. & Osokin, A.E. (1978) Propagation of one-dimensional waves through a nonlinear aftereffect medium, *Polymer Mechanics* **14:3**, 343-347.

Szemplinska-Stupnicka, W. & Bajkowski, J. (1986) The 1/2 subharmonic resonance and its transition to chaotic motion in a non-linear oscillator, *Int J Nonlin Mech* **21**, 401-419.

Tarasov, V.E. & Zaslavsky, G.M. (2006) Dynamics with low-level fractionality, *Physica A* **368**, 399-415.

Tavazoei, M.S. & Haeri, M. (2007) A necessary condition for double scroll attractor existence in fractional-order systems, *Phys Lett A* **367**, 102-113.

Tavazoei, M.S. & Haeri, M. (2008) Regular oscillations or chaos in a fractional order system with any effective dimension, *Nonlinear Dynam* **54**, 213-222.

Tavazoei, M.S. et al. (2009) More details on analysis of fractional-order van der Pol oscillator, *J Vib Control* **15**, 803-819.

Tarasov, V.E. (2006) Fractional variations for dynamical dynamical systems: Hamilton and Lagrange approaches, *J. Phys. A: Math. Gen.* **39**, 8409-8425.

Thabet, M. et al. (2008) Existence and uniqueness theorem for a class of delay differential equations with left and right Caputo fractional derivatives, *Journal of Mathematics and Physics* **49**, 083507.

Thai, H.T. & Kim, S.E. (2011) Nonlinear inelastic time-history analysis of truss structures, *Journal of Constructional Steel Research* **67**, 1966-1972.

Tien, W.M. et al. (1994) Nonlinear Dynamics of a Shallow Arch under Periodic Excitation .1. 1/2 Internal Resonance, *Int J Nonlin Mech* **29**, 349-366.

Tien, W.M. et al. (1994) Nonlinear Dynamics of a Shallow Arch under Periodic Excitation .2. 1/1 Internal Resonance, *Int J Nonlin Mech* **29**, 367-386.

Torvik, P.J. & Bagley, R.L. (1984) On the appearance of the fractional derivative in the behavior of real materials, *J Appl Mech–T Asme* **51**, 294-298.

Tsuda, Y. et al. (1992) Chaotic behavior of a nonlinear vibrating system with a retarded argument, *JSME International Journal, Series III* **35:2**, 259-267.

Uchaikin, V.V. (2013) *Fractional Derivatives for Physicists and Engineers*, Springer.

Urabe, M. (1965) Galerkin's procedure for nonlinear periodic systems, *Arch Ration Mech An* **20:2**, 120-152.

Verhulst, F. & Sanders, J.A. (1985) *Averaging Methods in Nonlinear Dynamical Systems* (Springer-Verlag, New York).

Wahi, P. & Chatterjee, A. (2004) Averaging oscillations with small fractional damping and delayed terms, *Nonlinear Dynam* **38**, 3-22.

Wang, Y.H. & Li, C.P. (2007) Does the fractional Brusselator with efficient dimension less than 1 have a limit cycle?, *Phys Lett A* **363**, 414-419.

Wang, Z.H. et al. (2011) Stability test of fractional-delay systems via integration, *Sci China Phys Chem* **54**, 1839-1846.

Wang, Z.H. et al. (2011) Stability testing of fractional delay systems via integration, Science in China G: Physics, *Mechanics & Astronomy* **54:10**, 1839-1846.

Wang, Z.H. & Hu, H.Y. (2010) Stability of a linear oscillator with damping force of the fractional-order derivative, *Sci China Phys Chem* **53**, 345-352.

Weilbeer, M. (2005) *Efficient numerical methods for fractional differential equations and their analytical background.* PhD Thesis, Technical University of Braunschweig.

Wineman, A. (2009) Nonlinear Viscoelastic Solids - A Review, *Math Mech Solids* **14**, 300-366.

Wu, B. & Li, P. (2001) A new approach to nonlinear oscillations, *J Appl Mech–T Asme* **68**, 951-952.

Wu, J.L. (2009) A wavelet operational method for solving fractional partial differential equations numerically, *Appl Math Comput* **214**, 31-40.

Wu, T.M. (2005) A study of convergence on the Newton-homotopy continuation method,

Appl Math Comput **168**, 1169-1174.

Wu, T.M. (2006) Solving the nonlinear equations by the Newton-homotopy continuation method with adjustable auxiliary homotopy function, *Appl. Math. Comput.* **173**, 383-388.

Wu, Z.M. et al. (2006) Analysing chaos in fractional-order systems with the harmonic balance method, *Chinese Phys* **15**, 1201-1207.

Xie, F. & Lin, X.Y. (2009) Asymptotic solution of the van der Pol oscillator with small fractional damping, *Phys Scripta* **T136**.

Xu, J. & Chung, K.W. (2003) Effects of time delayed position feedback on a van der Pol Duffing oscillator, *Physica D: Nonlinear Phenomena* **180**, 17-39.

Xu, J. & Chung, K.W. (2009) A perturbation-incremental scheme for studying Hopf bifurcation in delayed differential systems, *Science in China Series E*: Technological Sciences **52:3**, 698-708.

Xu, M.Y. & Tan, W.C. (2006) Intermediate processes and critical phenomena: Theory, method and progress of fractional operators and their applications to modern mechanics derivative constitutive relations, *Science in China: Series G Physics, Mechanics & Astronomy* **49:3**, 257-272.

Yabuno, H. (1997) Bifurcation control of parametrically excited duffing system by a combined linear-plus-nonlinear feedback control, *Nonlinear Dynam* **12**, 263-274.

Yabuno, H. et al. (1999) Stabilization of 1/3-order subharmonic resonance using an autoparametric vibration absorber, *J Vib Acoust* **121**, 309-315.

Yang, F. & Zhu, K.Q. (2010) Can we obtain a fractional Lorenz system from a physical problem?, *Chinese Phys Lett* **27:12**.

Yang, T.Z. & Fang, B. (2013) Asymptotic analysis of an axially viscoelastic string constituted by a fractional differentiation law, *Int J Nonlin Mech* **49**, 170-174.

Yang, Y.B. & Wu, Y.S. (1997) Chaotic behaviors of a two-member truss, *J Vib Control* **3**, 103-118.

You, X.C. & Xu, H. (2011) Analytical approximations for the periodic motion of the

Duffing system with delayed feedback, *Numer Algorithms* **56**, 561-576.

Yu, C. & Gao, G.Z. (2005) Existence of fractional differential equations, *J Math Anal Appl* **310**, 26-29.

Yuan, L.X. & Agrawal, O.P. (2002) A numerical scheme for dynamic systems containing fractional derivatives, *J Vib Acoust* **124**, 321-324.

Yuste, S.B. & Lindenberg, K. (2001) Subdiffusion-limited A+A reactions, *Phys Rev Lett* **87:11**, 118301.

Zhang, W. et al. (2009) Dynamic behaviors of nonlinear fractional-order differential oscillator, *Journal of Mechanical Science and Technology* **23**, 1058-1064.

Zhang, W. & Shimizu, N. (1998) Numerical algorithm for dynamic problems involving fractional operators, *Jsme Int J C−Mech Sy* **41**, 364-370.

Zhang, W. & Shimizu, N. (1999) Damping properties of the viscoelastic material described by fractional Kelvin-Voigt model, *Jsme Int J C−Mech Sy* **42**, 1-9.

Zhang, W. & Shimizu, N. (2001) FE formulation for the viscoelastic body modeled by fractional constitutive law, *Acta Mech Sinica−Prc* **17**, 354-365.

Zhang, X.Y. (2008) Some results of linear fractional order time-delay system, *Appl Math Comput* **197**, 407-411.

Zhang, Y.L. & Luo, M.K. (2012) Fractional Rayleigh-Duffing-like system and its synchronization, *Nonlinear Dynam* **70**, 1173-1183.

Zhao, F.Q. & Wang, M.M. (2013) Vibration analysis of an axially moving viscoelastic beam, *Applied Mechanics and Materials* **268-270**, 1177-1181.